Lecture Notes in Computer Science   **14280**

The series Lecture Notes in Computer Science (LNCS), including its subseries Lecture Notes in Artificial Intelligence (LNAI) and Lecture Notes in Bioinformatics (LNBI), has established itself as a medium for the publication of new developments in computer science and information technology research, teaching, and education.

LNCS enjoys close cooperation with the computer science R & D community, the series counts many renowned academics among its volume editors and paper authors, and collaborates with prestigious societies. Its mission is to serve this international community by providing an invaluable service, mainly focused on the publication of conference and workshop proceedings and postproceedings. LNCS commenced publication in 1973.

Abdelkader Hameurlain · A Min Tjoa
Editors

Transactions on Large-Scale Data- and Knowledge-Centered Systems LV

 Springer

Editors-in-Chief

Abdelkader Hameurlain
Paul Sabatier University
Toulouse, France

A Min Tjoa
Technische Universität Wien
Vienna, Austria

ISSN 0302-9743   ISSN 1611-3349 (electronic)
Lecture Notes in Computer Science
ISSN 1869-1994   ISSN 2510-4942 (electronic)
Transactions on Large-Scale Data- and Knowledge-Centered Systems
ISBN 978-3-662-68099-5   ISBN 978-3-662-68100-8 (eBook)
https://doi.org/10.1007/978-3-662-68100-8

This Springer imprint is published by the registered company Springer-Verlag GmbH, DE,
part of Springer Nature.
The registered company address is: Heidelberger Platz 3, 14197 Berlin, Germany

Paper in this product is recyclable.

Preface

This volume contains five fully revised regular papers, covering a wide range of very hot topics in the fields of data-driven science life science, workflows, weak signals, online social networks, root cause analysis, detected anomalies, analysis of interplanetary file systems, concept weighting in knowledge graphs, and neural networks.

We would like to sincerely thank the editorial board for thoroughly refereeing the submitted papers and ensuring the high quality of this volume. In addition, we would like to express our wholehearted thanks to the team at Springer for their ready availability, the efficiency of their management and the very pleasant cooperation in the realization of the TLDKS journal volumes.

July 2023

Abdelkader Hameurlain
A Min Tjoa

Organization

Editors-in-Chief

Abdelkader Hameurlain   Paul Sabatier University, IRIT, France
A Min Tjoa   Technical University of Vienna, IFS, Austria

Editorial Board

Reza Akbarinia   Inria, France
Dagmar Auer   Johannes Kepler University Linz, Austria
Djamal Benslimane   University Lyon 1, France
Stéphane Bressan   National University of Singapore, Singapore
Mirel Cosulschi   University of Craiova, Romania
Johann Eder   Alpen Adria University of Klagenfurt, Austria
Anna Formica   National Research Council in Rome, Italy
Shahram Ghandeharizadeh   University of Southern California, USA
Anastasios Gounaris   Aristotle University of Thessaloniki, Greece
Sergio Ilarri   University of Zaragoza, Spain
Petar Jovanovic   Universitat Politècnica de Catalunya and
     BarcelonaTech, Spain
Aida Kamišalić Latifić   University of Maribor, Slovenia
Dieter Kranzlmüller   Ludwig-Maximilians-Universität München,
     Germany
Philippe Lamarre   INSA Lyon, France
Lenka Lhotská   Technical University of Prague, Czech Republic
Vladimir Marik   Technical University of Prague, Czech Republic
Jorge Martinez Gil   Software Competence Center Hagenberg, Austria
Franck Morvan   Paul Sabatier University, IRIT, France
Torben Bach Pedersen   Aalborg University, Denmark
Günther Pernul   University of Regensburg, Germany
Viera Rozinajova   Kempelen Institute of Intelligent Technologies,
     Slovakia
Soror Sahri   LIPADE, Université Paris Cité, France
Joseph Vella   University of Malta, Malta
Shaoyi Yin   Paul Sabatier University, IRIT, France
Feng "George" Yu   Youngstown State University, USA

Contents

Life Science Workflow Services (LifeSWS): Motivations and Architecture

Reza Akbarinia[1]([✉]), Christophe Botella[1], Alexis Joly[1], Florent Masseglia[1],
Marta Mattoso[2], Eduardo Ogasawara[3], Daniel de Oliveira[4], Esther Pacitti[1],
Fabio Porto[5], Christophe Pradal[1,7], Dennis Shasha[6], and Patrick Valduriez[1]

[1] Inria, Univ Montpellier, CNRS, LIRMM, Montpellier, France
reza.akbarinia@inria.fr
[2] Federal University of Rio de Janeiro, Rio de Janeiro, Brazil
[3] CEFET/RJ, Rio de Janeiro, Brazil
[4] Fluminense Federal University, Rio de Janeiro, Brazil
[5] LNCC, Petrópolis, Brazil
[6] New York University, New York, USA
[7] CIRAD, AGAP Institute, Univ Montpellier, INRAE, Institut Agro Montpellier,
Montpellier, France

Abstract. Data driven science requires manipulating large datasets coming from various data sources through complex workflows based on a variety of models and languages. With the increasing number of big data sources and models developed by different groups, it is hard to relate models and data and use them in unanticipated ways for specific data analysis. Current solutions are typically ad-hoc, specialized for particular data, models and workflow systems. In this paper, we focus on data driven life science and propose an open service-based architecture, Life Science Workflow Services (LifeSWS), which provides data analysis workflow services for life sciences. We illustrate our motivations and rationale for the architecture with real use cases from life science.

Keywords: Data driven science · Life science · Data science · Workflows · Model life cycle · Service-based architecture

1 Introduction

Data driven science such as agronomy, astronomy, environmental, and life science must deal with overwhelming amounts of complex data, e.g., coming from sensors and scientific instruments, or produced by simulation. Increasingly, scientific breakthroughs will be enabled by advanced techniques from data science [23] that help researchers manipulate and explore these massive datasets [14].

Life science is the study of living organisms (plants, humans, microorganisms, . . .) and their association with internal or external conditions. It is an interdisciplinary domain including agronomy, biology, and botany. The data

© The Author(s), under exclusive license to Springer-Verlag GmbH, DE, part of Springer Nature 2023
A. Hameurlain and A. M. Tjoa (Eds.): *Transactions on Large-Scale Data-and Knowledge-Centered Systems LV*, LNCS 14280, pp. 1–24, 2023.
https://doi.org/10.1007/978-3-662-68100-8_1

in life science comes from many different data sources produced by modern platforms, e.g., high-throughput phenotyping, next-generation sequencing, remote sensing, etc., or readily available as international databases, e.g., Data.World, GenomeHub, AgMIP, EMPHASIS, etc. Such data is used to help producing/training models (statistical models, machine learning (ML) models, etc.) to derive information and knowledge or to make predictions using complex workflows. Since models are tailored to specific research questions, they are typically produced by different research groups and take various forms that reflect the researchers' approaches with their data.

Data processing with models typically involves complex data analysis workflows (*workflows*, for short hereafter). Unlike business workflow systems, e.g., new order processing, these workflows are compute- and data-intensive, may take hours or even days, but are often deterministic, and do not involve fine-grained transactions. They allow domain scientists (specialized in a science domain, e.g., plant biology), to express multi-step computational activities, such as loading input data files, processing the data, running analyses, and aggregating the results. Workflows have been implemented on top of scientific workflow systems such as Galaxy [1] and OpenAlea [30]. They frequently make use of data analytics engines such as Spark [36] and Flink [6], as well as Machine Learning (ML) libraries such as PyTorch [25] and Scikit-learn [26]. In order to scale to massive datasets, they make increasing use of distributed and parallel execution environments in the cloud.

While this paper (and project) focuses on data-driven life science, we believe the project can provide a framework for other application domains with similar requirements.

1.1 Use Cases

Let us illustrate the requirements for managing such data and models with real application examples from life science. In the context of climate change, agro-ecosystems face multiple challenges, including adaptation, resilience, epidemics, land-use conflicts, and the need for biodiversity conservation. Examples of practical questions that end-users might ask are:

- How to select or breed new plant varieties that are adapted to my local environmental conditions (e.g., drought, flooding, high temperature, disease)?
- Which treatments should be deployed on my farm depending on climatic conditions and geographical proximity to disease hot spots?

Addressing these questions requires multiscale modeling, e.g., modeling plants at different scales (e.g., organ, plant, crop, land surface, region) to predict the impact from heterogeneous data, e.g., data on plants, environment (weather, soil), and remote sensing. These models are the outcome of workflows, whose activities typically involve data extraction, data cleaning, machine learning, and visualization. Often the output of one workflow is the input to another.

With One Health, an approach that recognizes that the health of people is closely connected to the health of animals and our shared environment, understanding epidemic propagation at various levels (local, regional, national, global) has become critical for health authorities. The major problem is how to select the best prediction model for a given region by combining propagation models from different regions as well as integrating various data sources (epidemic, climate, socio-economic, etc.) along some common dimensions, e.g., time, location, etc.

The practical difficulty to achieve such integration is that it is hard to relate models and data, which are typically produced by different people with different methods, formats and tools. International repositories for scientific data and models are useful but they tend to be specialized for specific purposes and research communities, e.g., genomics, phenotyping, and epidemiology. Similarly, the workflow systems to manipulate data and models are specialized for a research domain, e.g., OpenAlea for plant phenotyping, Galaxy for genomics. Thus, there is a pressing need for integrated data and model management in order to achieve consistency and ease of use through generic workflow services with the ultimate goal of improving model accuracy and predictions.

1.2   The Centrality of Workflows

In this paper, we propose an open service-based architecture, called Life Science Workflow Services (LifeSWS). The main objective of LifeSWS is to help managing complex workflows by organizing massive and heterogeneous data, in connection with models and making workflow artifacts (datasets, models, metadata, workflow components, etc.) easy to search, debug, and parallelize.

In many ways, workflows are to scientific data processing what queries are to business data processing. In business data processing, queries must be written (with some reuse of other queries), debugged and optimized (sometimes through parallelization), and should work across distributed servers, hardware, and operating systems. Scientific data processing is much more complex so workflows replace queries. The issues however are much the same. Workflows in scientific data processing must be written (with some reuse of other workflow components), debugged (often benefiting from provenance), optimized (often through parallelization and caching), and should work across distributed servers and operating systems. In addition, workflows should be fault tolerant and workflow component versions should be kept up-to-date. Thus, a technical goal of this project is to make workflows work as seamlessly with data as queries do in business processing.

LifeSWS capitalizes on our previous experience in developing major systems for scientific applications such as: polystores with CloudMdSQL [17], workflows with OpenAlea [30], model management with Gypscie [35] [38], querying data across distributed services with DfAnalyzer [32] and Provlake [33], monitoring and debugging applications implemented in big data frameworks such as Apache Spark [12], and debugging workflows with BugDoc [18] and VersionClimber [29].

1.3   Paper Outline

The paper is organized as follows. Section 2 develops our motivating examples from real life science applications. Section 3 presents our open, service-based architecture for LifeSWS. Section 4 discusses platforms and infrastructures that can implement LifeSWS. Section 5 shows the use of LifeSWS with use cases from our motivating examples. Section 6 discusses related work. Section 7 concludes and discusses open research issues.

2   Motivating Examples

In this section, we introduce examples from real-life science applications that will serve as motivation for our work and as the basis for use cases with LifeSWS. These examples are in agro-ecosystems in the context of climate change and epidemic modeling. These examples share common requirements but have specific features that will show different uses of LifeSWS.

2.1   High-Throughput Phenotyping in the Context of Climate Change

To enhance the resilience of agro-ecosystems, interdisciplinary efforts are required, ranging from a detailed biological understanding of the physiology of plants with multiple stresses (e.g., drought, temperature, decease), agronomy to adapt agro-ecosystems to future challenges, as well as sociology, economy and politics to understand the impacts of changing public policy.

This challenge requires mobilizing all possible levers of plant adaptation, including the genotype, phenotype, and their interactions with the environment. The genetic/genomic revolution has allowed us to sequence and manipulate genes at a low cost and to generate a deluge of data. But understanding genome-to-phenotype relationships is crucial. While national and international phenotyping platforms allow the capture of phenotypes at high-throughput, most of the traits that contribute to the performance of agro-ecosystems are environment-dependent. The performance of a variety that thrives in a particular environment may perform poorly in a different one. Thus, it is important to capture phenotypes in various environments using various sensors at different scales (IoT, images from drones, 3D point clouds from Lidars and remote sensing images from satellites).

Major efforts have been invested in crop breeding to improve crop yield for food security. However, profiling the crop phenome by considering the structure and function of plants associated with genetics and environments remains a technical challenge [34].

In the past decade, high-throughput phenotyping platforms have emerged, enabling the collection of quantitative data on thousands of plants under controlled environmental conditions. A good example is the French Phenome project, with seven facilities producing 200 Terabytes of diverse, multiscale data annually, including images, environmental conditions, and sensor outputs from different sites [13].

To support high throughput phenotyping, many workflows have been developed using OpenAlea to analyze, reconstruct, and visualize the spatial and temporal development of the geometry and topology of thousands of plants in various environmental conditions. For instance, the *Phenomenal* workflow supports the reconstruction in 3D and the segmentation of plant organs [2]. The *PhenoTrack* workflow, which is based on Phenomenal, allows the 3D reconstruction of plants with the temporal tracking of the growth of each organ for the entire developmental cycle [9]. Finally, *RootSystemTracker* provides a workflow for the automatic structural and developmental 2D root phenotyping of Arabidopsis plants in Petri dishes [10].

Figure 1 shows a) the Phenomenal workflow implemented in OpenAlea; b) 3D organ tracking of a maize plant with PhenoTrack3D [9]; and c) a reconstructed root system architecture through time using RootSystemTracker [10].

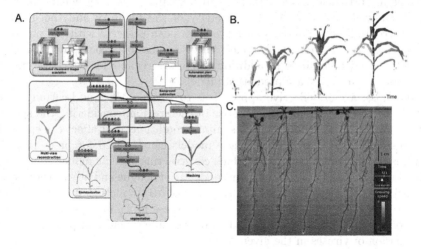

Fig. 1. Spatial and Temporal Workflows of Maize Shoot and Arabidopsis Root System Architecture

These workflows need to process large volumes of data on distributed infrastructures. To execute these workflows, we need to: 1) transfer large image datasets from a data center close to the phenotyping platforms to computing servers in the cloud; 2) distribute the execution on a cloud or grid infrastructure; 3) capture the provenance of the execution and cache intermediate results for later use; 4) rerun workflows with new processes and parameters; 5) provide execution results using dashboards to check the execution.

Furthermore, to understand the genotype-to-phenotype relationships, we need to be able to relate plant traits computed by phenotyping workflows (e.g., with OpenAlea) with genetic information using genotyping workflows such as genome-wide association studies (e.g., with Galaxy). Thus, we need to integrate heterogeneous workflows and be able to schedule their execution.

Through time, phenotyping workflows evolve with new processes implemented in various libraries whose versions and dependencies change quickly. Parameters need to be calibrated on new phenotyping platforms which have new sensors, different light conditions, or new plant species. Furthermore, it is important to identify problems in the workflow specification (e.g., that may lead to deadlocks) before executing them in HPC environments to avoid sparing resources. Finally, the workflows need to be debugged to identify problems occurring in new settings.

2.2   Epidemic Modeling

Each year, dengue, zika, chikungunya, and other arboviruses disseminated by the Aedes Aegypti vector exert an extreme burden on populations' health, especially in low-income countries.

General statistical models that try to explain or predict dengue in large areas usually do not consider the diversity of the territory and the different or even contradictory relations that predictors can preserve with the outcome. Conversely, creating individual models for every possible geographic location is impractical and unfeasible. In addition, there are regions for which we do not have enough data to create prediction models.

Therefore, a more effective approach is to develop Machine Learning models tailored to the unique characteristics of each region, considering the specific meteorological, socio-economic, and sanitary conditions that affect the epidemic transmission in that area. Then, these models can be used for predicting the transmission in similar regions for which constructing specific models is not possible (due to lack of data). For efficient modeling of dengue and other arboviruses, we need tools that can facilitate the selection of ML models that are most suitable for predicting these viruses. By utilizing these tools, more accurate predictive models can be selected and used to better understand and prepare for the transmission of viruses in the given region.

However, providing these tools is challenging. The reason is that we need to gather different datasets and models, and develop novel algorithms to enhance the accuracy and reliability of the prediction models. The required datasets and models are as follows: 1) *propagation datasets* that contain information about the spread of disease; 2) *climate datasets* that provide crucial insights into environmental factors like temperature and precipitation that may impact disease transmission; 3) *socio-economic datasets* that help us to understand the social and economic factors that could influence disease spread; 4) *prediction models* generated for some regions. By utilizing these diverse datasets and models, we should be able to select more accurate and reliable models for query regions, which can ultimately contribute to better disease control and prevention strategies.

Figure 2 shows the Dengue cases in Brazil spanning from 2000 to 2019 (a), along with the geographical distribution of cases across various regions (b and c) [5]. For predicting the Epidemy in each region, we need to select a model (or set of

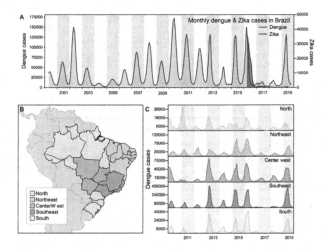

Fig. 2. Dengue Epidemy in Brazil

models) that takes into account the specific meteorological and socio-economic characteristics of the region.

3 Architecture of LifeSWS

In this section, we introduce the service-based architecture of LifeSWS, with its functional architecture and three layers of services (presentation and directory, workflow and data management services).

3.1 Functional Architecture

Our design choices are guided by the requirements of our users. The main potential users of LifeSWS are: the domain scientists who wish to analyze the data using different models and workflows; the workflow providers who create, maintain or enhance workflows for domain scientists using their workflow tools; the model providers who build models; and the data providers who supply data sources to the workflows.

Our architecture capitalizes on the latest advances in web-oriented architectures, microservices, containers and distributed and parallel data management [24]. We adopt the main following design choices and principles:

- Ease of use through web interfaces, which are easy to develop and specialize for different kinds of users;
- Open architecture with open source services and tools, and well-defined APIs to foster services interoperability (like cloud web services);
- Distributed architecture to provide performance, scalability and ease of use in the cloud using distributed database principles;

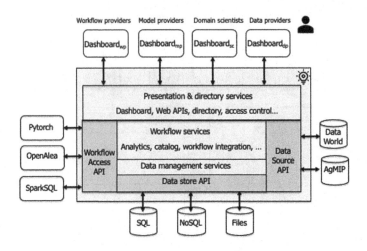

Fig. 3. LifeSWS Architecture

- Support for various databases (SQL, NoSQL, SciDB, etc.) and scientific file types (e.g., HDF and NetCDF);
- Integrated services, within the same data center or remote data centers and, with access to various tools and execution environments.

LifeSWS's functional architecture is shown in Fig. 3. It has three main layers of services: (1) presentation and directory, (2) workflow and (3) data management. Each layer can use the services of the same layer or the layers below. To interface with different systems, services can also use three kinds of APIs: Workflow Access APIs, Data Source APIs and Data Store APIs.

3.2 Presentation and Directory Services

Presentation and directory services provide users and applications with secure ways of accessing LifeSWS services. Presentation services include a Web dashboard service, a Web API and a directory service.

The Web dashboard service allows LifeSWS developers to build specific dashboards for different types of users (domain scientists, workflow providers, model providers and data providers). These dashboards allow users to analyze and display real-time data as charts and reports. They offer the following capabilities to developers: (1) a directory to publish and/or find data sources and workflow components; (2) tools for assembling workflows easily; (3) tools for debugging; and (4) scheduling workflows using workflow systems.

The directory stores data about LifeSWS users, access rights, dashboards and services. As a user directory, it helps register users, find out about them as well as authenticate them when accessing LifeSWS services. As a service directory, it provides a single place to publish, discover, and connect LifeSWS services as well as external services that can be distributed over the network. Additional network security, e.g., firewall, can be provided at this layer.

As an alternative to Web dashboards, the Web API is a server-side API that allows LifeSWS developers to access LifeSWS services from more general Web applications. This API consists of one or more publicly exposed endpoints that specify where and how to access the services with a request–response protocol, typically in JSON.

Finally, LifeSWS offers an external data view to ease the development of dashboards and workflows, which integrates observational and predictive data. This external view can be represented by a knowledge graph [15] extended to support the representation of observation time-series and predictive information metadata, such as: error estimate, multi-class prediction probabilities and etc.

3.3   Workflow Services

Workflow services make it easy for scientists to develop, debug and optimize their workflows for doing their scientific experiments and data analyses. The services should also support the of sharing data, models and workflow components. Because the users want to be able to use their familiar tools (e.g., workflow systems such as Galaxy or OpenAlea) and data sources (e.g., Data World), this layer provides efficient services to register and manage data and models, and allow model execution using different tools and data sources. The primary services provided at this layer are: catalog (including version management), model management, workflow integration and data analytics.

Catalog. The catalog is the central place to find out about all artifacts and tools of interest for LifeSWS users: data sources, datasets, models, workflows and code libraries. Artifacts can be found outside LifeSWS and thus accessed through some API, or stored within LifeSWS for efficient reuse. Each artifact has associated metadata that describes it and allows access to it, either locally if it is stored in LifeSWS, or through its URI if it is an external resource (tool or data source). With the catalog, one may register artifacts, change them or provide a new version. The catalog also knows about tools (e.g., OpenAlea, Spark) and code libraries that implement models (e.g., Phenomenal workflow). Finally, the catalog comes with a search capability that allows users to navigate through the hierarchy of artifacts.

Model Management. Various types of models are used in life sciences. Data-driven machine learning models adopt a learning strategy that updates a set of weights that approximates a function to the behavior of the learned phenomenon given by the patterns extracted from the input data. Typical tasks executed by machine learning models include solving classification and numerical regression problems, which one may generalize as prediction tasks. Another relevant type of model extensively used in life science are mechanistic models, which refer to computational artifacts derived from the mathematical modeling of a phenomenon.

The product of running mechanistic models for a certain number of time steps is referred to as a phenomenon simulation. For instance, crop simulation models reproduce the main functions of plants such as the evolution of plant architecture, light interception, photosynthesis, and water/nitrogen balance in the crop and soil [21].

The management of such life science model artifacts requires model life cycle management and model deployment, using specific tools that can be accessed through LifeSWS. Through a unified view of different model artifacts (produced with different tools), LifeSWS can improve model selection and allow for model integration.

Model selection allows the user to easily search for model artifacts of interest so they can be used for reproduction or integration. Searching can be done based on different criteria such as scientific domain or subdomain, metadata, format, tools and keywords. This capability uses the catalog of artifacts.

The performance monitoring of models in operation by the model management service is important to assess prediction quality and point to model updates. In particular, if the input data distribution changes, models built on past historical data must be flagged so they can be updated. For machine learning models, a concept-drift component must detect variations in input data patterns and launch alerts for downstream model updates. The latter are processed according to application requirements. LifeSWS supports complete and automatic model retraining, using ML tools, or it can delegate the model update process to components that implement a more sophisticated update procedure, involving for instance a fast training of a simple surrogate model, while the main model is updated.

Model integration allows combining different models, possibly produced using different tools. It can take different forms, depending on the model types and the integration objective. In machine learning, model integration may take the form of an ensemble of models [37]. An ensemble considers a set of models aiming at the prediction of the same target. The integration process is modeled as a pipeline that runs each individual participant model, possibly across different tools, over the same input and combines the individual results into an integrated one, often using a linear combination of the results. Using the DJEnsemble method [27], ensembles can be computed automatically by a LifeSWS platform such as Gypscie (see Sect. 4), so that the selection of participant models follows a cost-based selection approach. Model integration takes a different form in mechanistic models as they typically use scientific workflows for simulating the phenomenon. For example, to visualize a 3D model of plant growth in a local environment, the 3D plant structure and biophysical models such as light interception, carbon allocation and water and mineral uptake can be simulated by a workflow modeled in OpenAlea.

Workflow Integration. This service provides support for integrating and efficiently executing workflows on different workflow systems using the Workflow Access APIs. It shares some similar goals and functions found in data integration.

The main functions provided by this service are workflow definition and execution, provenance and cache.

For workflow definition and execution, we plan to rely on the Common Workflow Language (CWL) [8], an open standard aiming to enable scientists to share complex data analysis and machine learning workflows. CWL supports connecting command line tools to create workflows that are portable across a variety of CWL-compliant platforms, from a single developer's laptop up to a massively parallel cluster in the cloud. The CWL project produces free and open standards for describing command-line tool based workflows. These standards are implemented in many popular workflow systems such as Galaxy, Pegasus, Streamflow, and CWL-Airflow. To enable portability and reusability, CWL is explicit about inputs/outputs to form the workflow, data locations and execution models, which can be deployed using software container technologies, such as Docker and Singularity.

Within the CWL project, we can contribute to the definition of integrated workflows that span multiple workflows and workflow systems. Once an integrated workflow has been defined and its mappings registered using CWL, it can be executed using a LifeSWS scheduler that orchestrates execution across different workflow systems, in connection with these systems' schedulers.

Provenance (also referred to as lineage) management helps to reproduce, trace, assess, understand, and explain how datasets have been produced. This is a useful underlying functionality for several strategic capabilities, including experimental reproducibility, user steering (i.e., runtime monitoring, interactive data analysis, runtime fine-tuning) and data analysis. These capabilities are essential building blocks towards the goal of storing and sharing results of executions that can be useful later (by the same or different users perhaps on very different platforms).

In addition to provenance management, this service includes cache management, using information about cache data, as well as the location of the cache data (e.g., files, Spark RDDs, . . .). Caching datasets improves performance when they are produced at multiple times by different users or distributed at various sites. The decision whether to cache an intermediate result can be explicit (i.e., decided by the user) or made automatic based on workflow fragment analysis [13].

All information for this service is stored in a database that is relatively small. In particular, the cache itself is small (only references) and the cached data can be managed using the underlying execution environments accessed through the Workflow Access APIs. Using a database for this service provides the traditional advantages of data sharing, integrity and querying using an SQL-like language.

Analytics. This service allows scientists, through their specific dashboards, to perform analytics on the data produced by their workflows using the other workflow services. Using the Catalog and Model Management services, the user is able to select models and datasets of interest, execute the selected models using various workflow execution environments (using the Workflow Access APIs), and analyze the results. It also allows the user to cache and explain the results, and reproduce executions using the Provenance and Cache services.

This service also facilitates the analysis of different types of data such as time series and spatial data, by incorporating advanced analytical techniques like anomaly detection, similarity search and clustering.

3.4   Data Management Services

These services make it easy for LifeSWS users to manage their artifacts (datasets, models, metadata, etc.) and session data (logs, intermediate datasets, etc.) with high-level capabilities using the Data Source, Model/Workflow and Data Store APIs. An important capability is moving data between different data sources, databases and execution environments using simple import-export functions. Another useful capability, using the Data Source APIs, is subscribing to some data sources that provide a publish API, to get warned of the new versions.

More advanced capabilities, similar to distributed databases [24] and polystores [3], could be provided at this level for integrating data from different data sources. In particular, the CloudMdsQL polystore [17] is efficient for querying multiple heterogeneous data sources (e.g. files, relational and NoSQL) in the cloud. A CloudMdsQL query may contain nested subqueries, and each subquery addresses directly a particular data store and may contain embedded invocations to the data store native query interface. Thus, the major innovation is that a CloudMdsQL query can exploit the full power of local data stores, by simply allowing some local data store native queries to be called as functions, and at the same time be optimized based on a simple cost model, CloudMdsQL can also access address distributed processing frameworks such as Apache Spark by enabling the ad-hoc usage of user defined map/filter/reduce operators as subqueries.

Data Store APIs. These APIs allow storing and accessing data in different data stores (SQL, NoSQL, Savime, files, . . .), to support specific requirements. For instance, the catalog, provenance, and cache databases are typically in an SQL database such as PostgreSQL. By contrast, external data could be stored in the original format, e.g., JSON in a NoSQL document database, and extracted using data-specific APIs. Datasets produced using tools or cache data could be stored in files, e.g., Parquet, etc., or in a scientific database like Savime. The data store APIs should be based on standard APIs, such as JDBC, file system APIs, . . .).

Data Source APIs. These APIs allow connecting to various web data sources, such as Data World, AgMIP, etc., and performing various tasks (search for datasets, extract a dataset's metadata, import a dataset, get changes, etc.). They can be used to build user-friendly dashboards for domain scientists with semantic-based search capabilities, as for instance in the ontology-driven Phenotyping Hybrid Information System (PHIS [22]).

Workflow Access APIs. These APIs allow accessing and manipulating models and workflows as structured objects with their own semantics, execute them in their own execution environments, such as Pytorch (ML models), OpenAlea (workflows) and Spark (e.g., SparkSQL queries). These APIs make it possible to simply perform various tasks using models and workflows, such as import or export of models, executing them using a dataset, saving the result data (using the Data Store API, ...).

4 LifeSWS Platforms

LifeSWS services can be implemented and deployed in various platforms (using different software and hardware infrastructures) to address the specific requirements of vertical applications. Examples of platforms would be some LifeSWS services deployed in the cloud (public, private or hybrid) or on-premise clusters of servers, reusing existing software components that (partially) implement the services.

A good example of a LifeSWS platform is Gypscie [38], which provides services to develop, share, improve and publish scientific artifacts (datasets, models, etc.). Gypscie's services are available through two different interfaces. Figure 4 shows the interface that enables interactive access to services, including artifacts registration and service requests. The same functionality is available through a REST API based on the HTTP protocol.

These services make it easy for model providers and scientists to:

- Collect, curate and integrate heterogeneous data;
- Support the complete ML model life cycle, from model building to deployment, monitoring and policy enforcement;

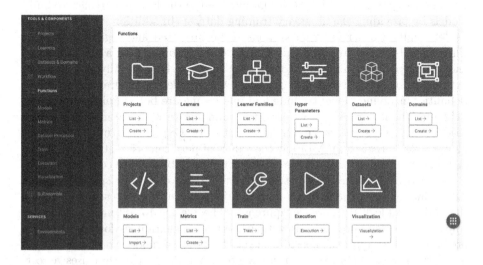

Fig. 4. Gypscie Web Interface

- Find ready-to-use models that best fit a particular prediction problem;
- Compare and ensemble models;
- Execute models with various tools: ML engines, workflow systems, ...
- Use specific hardware infrastructures and corresponding algorithms according to a desired task, e.g., use a distributed training algorithm for a particular GPU-based server for training a large deep neural network.

Let us illustrate how LifeSWS services would be supported by Gypscie services for presentation & directory, model management, and data management.

4.1   Presentation and Directory

Gypscie offers a web interface that eases ML model management. It also offers a notebook interface for direct python scripts integration with the Flask framework. Furthermore, Gypscie enables other tools to access its services through a RESTful-based API. Users can build dataflows graphically to model data preprocessing tasks. Registered dataflows can be scheduled for execution and, during runtime, have their activities and involved data recorded for provenance. User dataflows typically include data pre-processing transformations in preparation for model training and model inference.

4.2   Model Management

The core functionalities of Gypscie cover the services needed to support the full ML life cycle. Regarding model management, the Gypscie data model fosters the reuse of all artifacts involved during the model's life cycle. As such, the user can publish the scripts involving the data preparation and model fitting for a particular learning algorithm (hereafter, denoted *learner*). For ease of browsing, learners are aggregated into learner family. We use the learning artifact to build models by providing the necessary training dataset. In addition, Gypscie allows models built in other external systems to be imported and registered into it. Thus, models can be automatically registered when built using a known learner and the Gypscie training functionality, or they can be manually imported. In both cases, once registered they are ready to be called for inference. The functionalities involving model training and inferencing are both implemented using *MLFlow*. The latter enables Gypscie to communicate with the most frequently used ML engines, such as Pytorch and Scikit-learn. Gypscie instruments the running scripts to register in MLFlow the values of performance metrics, which the system collects and stores into its catalog, once the job has finished its execution.

A particular feature of Gypscie is its ability to deal with spatio-temporal data, which are extremely common in scientific applications. Gypscie implements the DJEnsemble [27] inference approach. The idea is to automatically select trained spatio-temporal models with performance guarantees for the scientific predictions. The approach considers a set of registered models for the execution of a certain task, for instance, rainfall prediction. The algorithm uses a cost-based strategy that strikes a balance between prediction precision and execution cost to select the best set of models that infer the rainfall prediction in a region of the space. Additionally, it specifies how to spatially allocate the selected models to cover the query region. Gypscie runs the optimization process, executes the selected models, and composes the final result. This is a very complex task that is completely abstracted from the final user, showing the potential of LifeSWS to create an easy ML environment.

4.3   Data Management

Data management involves the following services: (1) accessing registered data; (2) gathering provenance information, and (3) exploring the content of datasets. Gypscie registers metadata in its catalog for accessing data stored in an external data source, such as Databricks Delta Lake or Lustre file system. When a scheduled workflow requires a dataset, the dataset is automatically transferred from its stored location to a file system supporting the workflow execution environment.

As a workflow applies transformations on a dataset, Gypscie stores the provenance information regarding the operation. Thus, within Gypscie a user can always review the lineage of transformation that led to the dataset's current version.

Finally, Gypscie integrates the SAVIME in-memory multi-dimensional array database system [20]. SAVIME supports the expression of SQL-like queries over raw datasets. The query language enables the registration of prebuilt ML models that can be invoked over the results of a query expression.

Gypscie has been deployed on a server at LNCC, and interfaced with two execution environments (see Sect. 5: the Santos Dumont HPC system at LNCC which could be used in large model training (e.g., using PyTorch) and a Spark shared-nothing cluster to perform large-scale data transformation.

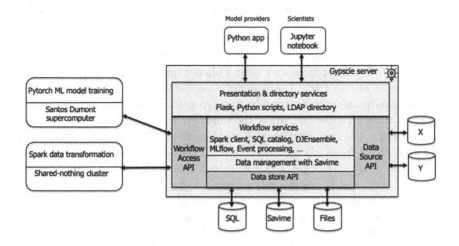

Fig. 5. LifeSWS platform with Gypscie

5   Use Cases with LifeSWS

In this section, we show how the LifeSWS services could be used to support the requirements of the two motivating examples, referring to the services of the previous section.

5.1   High-Throughput Phenotyping

In this use case, LifeSWS is used by domain scientists in order to analyze, process and visualise High Throughput Phenotyping (HTP) experiments, with the following workflows, models and datasets:

- OpenAlea workflows that implement a specific phenotyping processing such as 3D maize segmentation, organ tracking, or root system reconstruction;
- A Galaxy workflow for a Genome-Wide Association Study (GWAS);
- Image analysis algorithms to segment the background, to reconstruct the plant in 3D using space carving, semantic segmentation and tracking of the organs;
- Functional-structural plant models that are used either to compute non-observed information like light interception by leaves [2] or water fluxes inside the root system [4], or to generate synthetic data for training or validating methods of plant reconstruction;
- Raw image datasets obtained from the phenotyping platform, which contain timeseries of several images per plant;
- Outputs of plant traits (e.g., leaf angle, light use efficiency, or biomass accumulation) for each genotype are saved in the Phenotyping Hybrid Information System (PHIS) [22].

A domain scientist, often a plant biologist, searches for a specific OpenAlea phenotyping workflow based on its metadata. Then, she edits, visualizes, and executes the workflow on a small dataset. For instance she can reconstruct in 3D the growth and development of a maize plant during a growing season. Finally, she selects a full dataset from an existing phenotyping experiment and executes the workflow to obtain a set of plants traits specific to each genotype and to environmental conditions such as drought or temperature. Finally, to breed varieties tolerant to drought, some outputs of the workflow will be associated with genetic markers using a Galaxy workflow that implements a Genome-Wide Association Study (GWAS). GWAS allows to correlate favorable traits (e.g., responsible for drought tolerance) with a genomic region and thus to breed new varieties.

For instance, the Phenomenal workflow (Fig. 1.a) is composed of different fragments, i.e. reusable sub-workflows: binarization, images calibration, 3D volume reconstruction, and organ segmentation. The intermediate datasets are also shown in the Figure. The raw data is produced by the Phenoarch platform, which has a capacity of managing 2500 plants within a controlled environment (e.g., temperature, humidity, irrigation) and automatic imaging through time. The total size of the raw image dataset for one experiment is 15 Terabytes. The raw data is stored on a server close to the experimental platform, but also referenced to the PHIS with all metadata.

To execute the Phenomenal workflow, the OpenAlea scheduler checks, using provenance data, whether some fragments have already been executed and are present in the cache. OpenAlea then schedules the execution of the workflow on a distributed infrastructure. After execution, the user can visualize the results either as classical plots or as 3D plots to inspect the reconstructed plants. The results are automatically stored in the PHIS, to associate each plant of each genotype with environmental data and the computed traits. These results are used as input of the Galaxy workflow to make a complete GWAS study.

Let us now explain how this use case can be realized using the services provided by LifeSWS. Required workflows of the use case can be searched and found using the Catalog service. This allows to navigate among OpenAlea and Galaxy workflows and to select the Phenomenal and GWAS workflows. Both workflows are composed of versioned tools, models, and workflow fragments that are retrieved from the Model Management service. Workflows are visualized and parameters are set via workflow dashboards of the Presentation and Directory services. After edition, new versions are stored using the Model Management service. Datasets of the Phenomenal experiment can be retrieved and accessed using the Data Source APIs with a connection to the PHIS.

LifeSWS looks up the provenance and cache, and triggers OpenAlea's distributed execution. Provenance supports the determination of whether a workflow fragment has been already computed with the same parameters and datasets. The cache enables the retrieval of previous intermediate results rather than recomputing them again. The cache and provenance information are updated during the execution of OpenAlea workflows using the Workflow Access API. LifeSWS provides Data Services to feed the Galaxy workflow with the output of OpenAlea workflow when execution has been done. Then, it triggers the execution of Galaxy on distributed resources.

The visualization of the intermediate and final results is done in a workflow dashboard and some specific outputs (e.g., plant traits) are updated in the PHIS using the Data Source APIs.

Moreover, OpenAlea models and workflows can be upgraded automatically using VersionClimber [29] from the Workflow Integration services to find the latest compatible versions of all the modela and libraries the workflows depend on.

5.2 Epidemic Modeling

In this use case, LifeSWS is used by scientists in order to select ML models that would perform best to predict the transmission of the Dengue virus in a particular region.

In this scenario, the following inputs may be used by LifeSWS:

– Propagation datasets obtained from public health institutions, which are likely to contain information on the spread of disease.

– Publicly available climate datasets, which could be used to identify environmental factors, such as temperature and precipitation, that may impact disease transmission.
– Socio-economic datasets publicly available for specific countries and regions, which may provide insight into social and economic factors that could influence the spread of disease.
– Prediction models provided by the system's users for different regions.

Let r be a region in a country given by a user, the objective is to find the best models that can predict the dengue transmission for r. If there are available models for the given region, then they are returned to the user.

Otherwise, the system should execute the following workflow activities to analyze the characteristics of region r, and find the appropriate prediction models. In this case, the system performs a similarity search between the variables representing r (e.g., meteorological and socio-economical variables) and those of the regions for which it has predictive models in its database. Using this similarity search, LifeSWS identifies the most similar region to r. Finally, it retrieves the predictive models associated with the most similar region and returns them as the best models for predicting Dengue transmission in r.

Let us now explain how this use case can be realized using the services provided by LifeSWS. The required datasets of the use case, mainly propagation, climate and socio-economic datasets can be accessed using the Data Source APIs. The prediction models are given to LifeSWS via the dashboards of the Presentation and Directory services. Then, they are stored using the Model Management service. The Catalog is used to index the metadata of the given datasets and models.

The whole workflow for finding the prediction models of the given region is executed by the scheduler of the Workflow Integration services. The users send their region r to LifeSWS via their dashboard. Then, the system checks the Catalog to determine whether there are any predicting models for the given region. If the answer is positive, the models are retrieved from the Model Management service. Otherwise, LifeSWS needs to find the similar regions to r and return their predicting models. For this, LifeSWS first uses the Data Store service, to find the metadata of the query r and other regions. Then, the Data Analytics service is used to perform a similarity search in order to find the most similar regions to r, which we denote as R. LifeSWS uses its Catalog to select the best models for predicting the transmission of the Dengue virus in the regions similar to r. Then, it calls the Cache management service to retrieve the selected models if they are available in the cache. Otherwise, the Model Management service is used to access the selected models, which are then returned to the user using the dashboards.

6   Related Work

To address our objectives, many different approaches and solutions could be used with different trade-offs between development and maintenance cost, generality and efficiency. In this section, we discuss the main practical approaches and related technologies in a wide spectrum from generic to specific: cloud services, scientific workflow systems, heterogeneous data management systems, model life-cycle frameworks, and science platforms.

At one end of the spectrum (the most generic approaches), we have cloud services from major vendors (Amazon, Google, Microsoft, IBM, Oracle, . . .). They provide many ready-to-use services within a Platform-as-a-Service (PaaS) to build applications that deal with Web data and enterprise data. They focus on ease-of-use, elasticity and interoperability through well-defined APIs that allow to use proprietary as well as open-source software. For instance, Amazon Web Services (AWS) is a large cloud computing platform, offering 200+ services, from basic services (storage, computing, database, containers, security, . . .) to more advanced services (machine learning, data warehouse, data lake, search, . . .). However, a first reason that prevents the use of such cloud platforms for LifeSWS is the lack of services directly available for scientific applications (work-flows, provenance, numerical simulation, interface to HPC systems, . . .). Another important reason for scientific organizations is that they prefer to rely on open vendor-neutral vendors.

At the other end of the spectrum (the most specific approaches), we have scientific workflows systems, such as Galaxy [1], Kepler [19] and OpenAlea [30], which are designed to help scientists developing complex applications. They typically include tools to model, design, debug, share and execute workflows, with interactive visualization of the results. To support result analysis and explaining and experiment reproducibility, workflow systems often support provenance, which captures the derivation history of a dataset, including the original data sources, intermediate datasets, and the computational steps that were applied to produce this dataset. Workflows are also often data-intensive, i.e., process, manage or produce huge amounts of data. Thus, in order to be executed in reasonable time, they require deployment in High Performance Computing (HPC) environments such as supercomputers, computer clusters or grids. For instance, DfAnalyzer [32] is a tool that enables monitoring, debugging, steering, and analysis of dataflows while the data is being generated by scientific applications. Most workflow systems are also open-source, providing access to community shared resources such as models, code libraries, and datasets. Thus, they tend to be specialized for some scientific domains. For instance, Galaxy is quite popular in bioinformatics, while OpenAlea is specialized in plant phenotyping. Kepler [19] addresses other scientific domains such as chemistry, ecology, geology, molecular biology and oceanography.

More generic, we have the popular big data analytics engines such as Spark [36] and Flink [6] which allow for batch or realtime data processing, and ML libraries such as PyTorch [25] and Scikit-learn [26] with workflows to collect training data, preprocess data (cleaning, formatting, . . .), build datasets, train and refine models and evaluate.

As workflows are getting used a lot in practice, the problem of debugging has become important. It is difficult since there are many potential sources of errors including: bugs in the code, input data, software updates, and improper parameter settings. To address this problem, BugDoc [18] automatically infers the root causes and derive succinct explanations of failures for black-box pipelines using the results from previous runs. VersionClimber [29] is another automated system that deals with the problem of the pipelines that apply multiple packages, each of which evolves independently, to one or several data sources. VersionClimber automatically discovers newer versions of these packages that are compatible.

For the applications envisioned in LifeSWS, these systems will help, because we may want to combine different workflows (e.g., Galaxy, OpenAlea and Spark), debug them with a tool like BugDoc with some other data analytics services (e.g., time series analysis) and keep versions up-to-date. To integrate and execute heterogeneous workflows, we plan to rely on the Common Workflow Language (CWL) standard [8], which helps creating portable workflows.

Heterogeneous data management systems provide capabilities to access heterogeneous different data sources, which are important for our objectives. The problem of querying heterogeneous data sources, i.e., managed by different data management systems such as relational or XML database systems, has long been studied in the context of multidatabase systems [24]. However, multidatabase systems have not been designed for the cloud, with a large variety of data stores such as SQL, NoSQL, NewSQL and HDFS. Furthermore, operating in a cloud infrastructure provides more control over where the system components can be installed, which makes it possible to design more efficient architectures. These differences have motivated the design of polystores (or multistore systems) that provide integrated access to a number of cloud data stores. For instance, CloudMdsQL [17] supports a functional SQL-like language, capable of querying multiple heterogeneous data stores within a single query that may contain embedded invocations to each data store's native query interface.

Spurred by the growing use of machine learning in all kinds of applications, many new model lifecycle systems have been proposed. Different from traditional software engineering, the development of ML applications is more iterative and explorative, yielding a variety of artifacts, such as datasets, models, features, hyperparameters, metrics, software code and pipelines. The objective of these new systems is to enable explainability, reproducibility, and traceability of ML executions by supporting the storage, management and reuse of these artifacts. The systematic literature review of more than 60 ML lifecycle management systems [31] shows that there is no precise functional scope, thus making comparison between systems difficult. Some systems focus on the management of ML artifacts only while some others add capabilities for the development of ML applications. The most complete systems come from cloud providers, e.g., Microsoft Azure ML, Amazon SageMaker and Google Vertex AI, as ML as a service (MLaaS) platforms. In contrast, open-source systems tend to be more focused. For instance, MLflow [7] focuses on capturing, storing, managing, and deploying ML artifacts using a standard format to store models and project code. It provides APIs to access ML development tools, such as PyTorch, Scikit-learn and Tensorflow. Also motivated by the objective of providing a holistic view to support the lifecycle of scientific ML, ProvLake [33] is a provenance data management system capable of capturing, integrating, and querying data across multiple distributed services, programs, databases, stores, and computational workflows by leveraging provenance data.

Science platforms are facilities that provide services and resources for research communities to perform collaborative research, observation and experimentation. They may include major scientific equipment, sometimes HPC machines, scientific datasets, data and research papers, code libraries and models. A common particular case is the science gateway (or science portal), which is a community-developed set of tools, applications, and data that are integrated through a web-based portal or a suite of applications. Science platforms are more or less specialized for some particular science, e.g., InfraPhenoGrid, PHIS, Plntnet and CyVerse.

InfraPhenoGrid [28] is a grid-based platform to efficiently manage datasets produced by the PhenoArch plant phenomics platform in Montpellier and deploy scientific workflows using a middleware that hides complexity.

PHIS [22] is a rich Phenotyping Hybrid Information System complementary to InfraPhenoGrid designed for plant phenomics. It allows storing and managing heterogeneous data (e.g., images, spectra, growth curves) and multi-spatial and temporal scale data (leaf to canopy level) coming from multiple sources (field, greenhouse). Its ontology-driven architecture is a powerful tool for integrating and managing data from multiple experiments and platforms including field and greenhouse. PHIS allows to enrich datasets with knowledge and metadata enabling the reuse of data and meta-analyses. In contrast, LifeSWS addresses a wider spectrum of applications in life sciences and provides key services such as user-specific Web dashboards, model management, provenance and cache, and workflow integration.

PlntNet [16] is a participatory platform and information system dedicated to the production and sharing of botanical data in order to study biodiversity. The main application performs deep learning-based plant identification on a smartphone and returns, given a plant image, the ranked list of the most likely species and asks for interactive validation by the users.

CyVerse [11] is a platform for life sciences with services and resources to deal with huge datasets and complex data analyses. It includes a Web-based platform with data management services (storage, analysis, visualization, exploration), shared data and science APIs to access supercomputing resources. CyVerse is the closest to LifeSWS, but lacks key services such as user-specific Web dashboards, model management, provenance and cache, and Model/Workflow APIs.

7   Conclusion

In this paper, we proposed LifeSWS, an open service-based architecture that implements data analysis workflow services for life sciences. The main objective of LifeSWS is to support the construction and maintenance of high quality, scalable and efficient workflows by organizing and making workflow artifacts (datasets, models, metadata, workflow components, etc.) easy to search and manipulate using various workflow systems.

Our architecture capitalizes on the latest advances in web-oriented architectures, microservices and distributed and parallel data management. It relies on open source services and tools, and well-defined APIs to foster services interoperability (like cloud web services). LifeSWS provides three main layers of services (presentation and directory, workflow and data management) and APIs to interface with different workflow systems, data sources and data stores.

LifeSWS services can be implemented and deployed in various platforms to address the specific requirements of vertical applications. We illustrated a LifeSWS platform with Gypscie, which provides services to develop, share, improve and publish scientific artifacts (datasets, models, etc.).

We also illustrated our proposed architecture with real use cases from life science. These examples are in agro-ecosystems in the context of climate change and epidemic modeling. These examples share common requirements but have specific features that show different uses of LifeSWS.

LifeSWS capitalizes on our previous experience in developing major systems for scientific applications. However, there are still major issues. To understand the research issues, let us consider two important scenarios in which LifeSDS should be able to help: (1) domain scientists build one or more datasets which may be in a variety of formats (relational database, csv files, etc.); (2) they also build workflows that make use of these datasets. LifeSDS comes to play a role in two major ways: (1) improve the maintenance and performance of existing workflows; (2) allow authenticated and efficient access and management of multiple workflows and datasets.

Based on our experience and application use cases, the main issues we plan to work on involve:

- Making it easy to integrate and run heterogeneous workflows defined using the Common Workflow Language (CWL), while providing reuse and reproducibility;
- Providing efficient execution of heterogeneous workflows by caching intermediate results and performing cache-aware scheduling;
- Making it easy for domain scientists to manage the model life cycle, perform model selection and model integration for different types of models managed using different tools;
- Assisting scientists in analyzing diverse data types, such as time series, through the integration of advanced methods such as clustering, anomaly detection, kNN search, etc., for different analytical requirements;
- Keeping track of the provenance of both data sources and software components, in order to aid in debugging using tools such as BugDoc [18] and to enhance the reproducibility of computational experiments.

Acknowledgement. This work is within the context of the HPDaSc associated team between Inria and Brazil. Some of us are supported by CNPq research productivity fellowships. C. Pradal has support from the MaCS4Plants CIRAD network, initiated from the AGAP Institute and AMAP joint research units, and EU's Horizon 2020 research and innovation program (IPM Decisions project No. 817617, BreedingValue project No. 101000747).

References

1. Afgan, E., et al.: The galaxy platform for accessible, reproducible and collaborative biomedical analyses: 2022 update. Nucleic Acids Res. **50**(W1), 345–351 (2022)
2. Artzet, S., et al.: Phenomenal: an automatic open source library for 3D shoot architecture reconstruction and analysis for image-based plant phenotyping. BioRxiv p. 805739 (2019)
3. Bondiombouy, C., Valduriez, P.: Query processing in multistore systems: an overview. Int. J. Cloud Comput. **5**(4), 309–346 (2016)
4. Boursiac, Y., et al.: Phenotyping and modeling of root hydraulic architecture reveal critical determinants of axial water transport. Plant Physiol. **190**(2), 1289–1306 (2022)
5. Brito, A., et al.: Lying in wait: the resurgence of dengue virus after the zika epidemic in Brazil. Nat. Commun. **12**, 2619 (2021)
6. Carbone, P., Katsifodimos, A., Ewen, S., Markl, V., Haridi, S., Tzoumas, K.: Apache flink: stream and batch processing in a single engine. IEEE Data Eng. Bull. **38**(4), 28–38 (2015)
7. Chen, A., et al.: Developments in MLflow: a system to accelerate the machine learning lifecycle. In: Workshop on Data Management for End-To-End Machine Learning (DEEM@SIGMOD), pp. 5:1–5:4 (2020)
8. Crusoe, M.R., et al.: Methods included: standardizing computational reuse and portability with the common workflow language. Commun. ACM **65**(6), 54–63 (2022)
9. Daviet, B., Fernandez, R., Cabrera-Bosquet, L., Pradal, C., Fournier, C.: Phenotrack3d: an automatic high-throughput phenotyping pipeline to track maize organs over time. Plant Methods **18**(1), 130 (2022)
10. Fernandez, R., Crabos, A., Maillard, M., Nacry, P., Pradal, C.: High-throughput and automatic structural and developmental root phenotyping on arabidopsis seedlings. Plant Methods **18**(1), 1–19 (2022)
11. Goff, S., et al.: The iplant collaborative: cyberinfrastructure for plant biology. Front. Plant Sci. **2** (2011)
12. Guedes, T., et al.: Capturing and analyzing provenance from spark-based scientific workflows with samba-rap. Future Gener. Comput. Syst. **112**, 658–669 (2020)
13. Heidsieck, G., de Oliveira, D., Pacitti, E., Pradal, C., Tardieu, F., Valduriez, P.: Cache-aware scheduling of scientific workflows in a multisite cloud. Futur. Gener. Comput. Syst. **122**, 172–186 (2021)
14. Hey, T., Tansley, S., Tolle, K., Gray, J.: The Fourth Paradigm: Data-Intensive Scientific Discovery. Microsoft Research, October 2009
15. Hogan, A., et al.: Knowledge graphs. ACM Comput. Surv. **54**(4) (2021). https://doi.org/10.1145/3447772
16. Joly, A., et al.: Interactive plant identification based on social image data. Ecol. Inform. **23**, 22–34 (2014). Special Issue on Multimedia in Ecology and Environment
17. Kolev, B., Bondiombouy, C., Valduriez, P., Jiménez-Peris, R., Pau, R., Pereira, J.: The CloudMdSQL multistore system. In: ACM SIGMOD International Conference on Management of Data, pp. 2113–2116 (2016)

18. Lourenço, R., Freire, J., Simon, E., Weber, G., Shasha, D.E.: Bugdoc: iterative debugging and explanation of pipeline. VLDB J. **32**(1), 75–101 (2023)
19. Ludäscher, B., et al.: Scientific workflow management and the Kepler system. Concurr. Comput. Pract. Exp. **18**(10), 1039–1065 (2006)
20. Lustosa, H.L.S., da Silva, A.C., da Silva, D.N.R., Valduriez, P., Porto, F.A.M.: SAVIME: an array DBMS for simulation analysis and ML models predictions. J. Inf. Data Manag. **11**(3), 247–264 (2021)
21. Muller, B., Martre, P.: Plant and crop simulation models: powerful tools to link physiology, genetics, and phenomics. J. Exp. Bot. **70**(9), 2339–2344 (2019)
22. Neveu, P., et al.: Dealing with multi-source and multi-scale information in plant phenomics: the ontology-driven phenotyping hybrid information system. New Phytol. **221**(1), 588–601 (2019)
23. Özsu, M.T.: Data science: a systematic treatment. Commun. ACM **66**(7), 106–116 (2023)
24. Özsu, M.T., Valduriez, P.: Principles of Distributed Database Systems, 4th edn. Springer, Cham (2020). https://doi.org/10.1007/978-3-030-26253-2
25. Paszke, A., et al.: Pytorch: an imperative style, high-performance deep learning library. In: Annual Conference on Neural Information Processing Systems (NeurIPS), pp. 8024–8035 (2019)
26. Pedregosa, F., et al.: Scikit-learn: machine learning in Python. J. Mach. Learn. Res. **12**, 2825–2830 (2011)
27. Pereira, R.S., et al.: Djensemble: a cost-based selection and allocation of a disjoint ensemble of spatio-temporal models. In: International Conference on Scientific and Statistical Database Management (SSDBM), pp. 226–231 (2021)
28. Pradal, C., et al.: InfraPhenoGrid: a scientific workflow infrastructure for Plant Phenomics on the Grid. Futur. Gener. Comput. Syst. **67**, 341–353 (2017)
29. Pradal, C., Cohen-Boulakia, S., Valduriez, P., Shasha, D.: VersionClimber: version upgrades without tears. IEEE Comput. Sci. Eng. **21**(5), 87–93 (2019)
30. Pradal, C., Fournier, C., Valduriez, P., Boulakia, S.C.: OpenAlea: scientific workflows combining data analysis and simulation. In: International Conference on Scientific and Statistical Database Management (SSDBM), pp. 11:1–11:6 (2015)
31. Schlegel, M., Sattler, K.: Management of machine learning lifecycle artifacts: a survey. ACM SIGMOD Rec. **51**(4), 18–35 (2022)
32. Silva, V., de Oliveira, D., Valduriez, P., Mattoso, M.: DfAnalyzer: runtime dataflow analysis of scientific applications using provenance. Proc. VLDB Endow. (PVLDB) **11**(12), 2082–2085 (2018)
33. Souza, R., et al.: Workflow provenance in the lifecycle of scientific machine learning. Concur. Comput. Pract. Exp. **34**(14) (2022)
34. Tardieu, F., Cabrera-Bosquet, L., Pridmore, T., Bennett, M.: Plant phenomics, from sensors to knowledge. Curr. Biol. **27**(15), R770–R783 (2017)
35. Valduriez, P., Porto, F.: Data and machine learning model management with Gypscie. In: CARLA workshop on HPC and data sciences meet scientific computing, pp. 1–2 (2022)
36. Zaharia, M., Chowdhury, M., Franklin, M.J., Shenker, S., Stoica, I.: Spark: cluster computing with working sets. In: USENIX Workshop on Hot Topics in Cloud Computing (HotCloud) (2010)
37. Zhang, C., Ma, Y.: Ensemble Machine Learning, Methods and Applications. Springer, New York (2012). https://doi.org/10.1007/978-1-4419-9326-7
38. Zorrilla, R., Ogasawara, E.S., Valduriez, P., Porto, F.: A data-driven model selection approach to spatio-temporal prediction. In: Brazilian Symposium on Databases (SBBD), pp. 1–12 (2022)

The Power of Weak Signals: A Twitter Analysis on Game of Thrones' Final Season

Hiba Abou Jamra and Marinette Savonnet[(✉)]

LIB - EA 7534, University of Burgundy,
9, Avenue Alain Savary, 21078 Dijon, France
{hiba.abou-jamra,marinette.savonnet}@u-bourgogne.fr

Abstract. Online Social Networks are built on relationships between individuals. Moreover, they are a space where particular events emerge more quickly than through traditional media such as newspapers or radio. These networks represent a good source for detecting indicators of changes, called weak signals, announcing future threats or opportunities for an organization. The identification of weak signals in a complex environment exposed to ongoing evolution is still a challenging task in research. This paper proposes an approach based on the topology of networks to identify weak signals. Our approach contrasts existing works that analyze significant themes and trends, i.e., strong signals prevalent in a social network at a particular time. We model social network data in the form of temporal interaction graphs. Then we choose graphlets (particular network motifs) as an operational tool to estimate the characteristics of weak signals. We demonstrate our approach accordingly by a study on retweets published in the last season of the TV series Game of Thrones, with which we evaluated and validated some properties of our approach.

Keywords: Weak signals · Online Social Networks · Network topology · Graphlets

1 Introduction and Motivations

Recently, the Web has become an essential tool for communication where knowledge-sharing platforms, like social networks, have expanded. Online Social Networks bring together millions of people and strengthen their relationships and exchanges with new forms of cooperation and communication. There are several types of Online Social Networks, such as generalist platforms like Twitter or Facebook, professional platforms like LinkedIn, and media-sharing platforms like Youtube, Flickr. Exploring these networks generally focuses on community detection, link prediction, user sentiment analysis, user classification, user profile construction to characterize their skills and knowledge, influence analysis, ... [48]. Trends or events emerge more quickly through Online Social Networks

© The Author(s), under exclusive license to Springer-Verlag GmbH, DE, part of Springer Nature 2023
A. Hameurlain and A. M. Tjoa (Eds.): *Transactions on Large-Scale Data-and Knowledge-Centered Systems LV*, LNCS 14280, pp. 25–63, 2023.
https://doi.org/10.1007/978-3-662-68100-8_2

than traditional media like websites, radio, television, or newspapers [1]. For example, Jane Jordan-Meier in [25] discusses Twitter's role in spreading information regarding the water landing of a plane on the Hudson River in January 2009. The first tweet, *"There's a plane in the Hudson. I'm on the ferry going to pick up the people. Crazy"* accompanied by a photo, is sent by Janis Krum to her 170 followers. Thirty-two minutes after this tweet, traditional media began to broadcast and report on the incident[1]. Twitter acts as an echo chamber in which users are exposed to consistent opinions with their views (also known as a bubble filter). This echo effect amplifies events, affecting public and industrial policies and user behavior.

Organizations have to make decisions in an environment subject to constant and rapid change. Indeed, the last few years have been marked by multiple health, economic and geo-political crises that have highlighted the importance of weak signals to help governments and companies understand and adapt to changes in their environment as early as possible [42,51]. Detecting and interpreting these weak signals when events are not known in advance are complex problems for researchers. Since weak signals are small precursors, they are usually hidden in the middle of a large amount of information. In addition, time constraints limit decision-making that requires rapid processing of a large volume of data [52]. In 1978, Turner estimated that most accidents are related to chains of errors and defects, which are almost always detectable before the accident occurs [38].

In this paper, we focus on identifying weak signals in Online Social Networks and more globally in graphs of temporal interactions between entities. We first explore the data of these environments to detect weak signals of change and then interpret them to make relevant and valuable decisions for the organization. To be able to respond to the research problem, we will present in this paper our BEAM[2] method, that we proposed for the detection and interpretation of weak signals in Online Social Networks.

The remainder of the paper is structured as follows: Sect. 2 introduces definitions of weak signals from the literature, and Sect. 3 discusses a few of the methods used to detect them. In Sect. 4, we present our proposed approach to detect and interpret weak signals in social networks, followed by a case study in Sect. 5. Section 6 presents two experiments to validate the approach. This series of studies is carried out on the final Game of Thrones season airing. Section 7 is a complementary proposal based on SVD to filter graphlets. Finally, Sect. 8 concludes the work presented in this paper and discusses future directions and perspectives.

---

[1] https://www.cnbc.com/2014/01/15/the-five-year-anniversary-of-twitters-defining-moment.html.

[2] BEAM: a ray or shaft of light beams from the searchlights, a collection of nearly parallel rays (such as X-rays) or a stream of particles (such as electrons), a constant directional radio signal transmitted for the guidance of pilots.

2   Weak Signal: A Notion with Many Facets

The weak signals concept has been developed in various fields, such as signal processing, information theory, business strategy, crisis management, and industrial risk prevention. Due to this diversity, weak signals have a rich lexical field. Several terms have emerged, such as "hunch" in crisis management, "alarm signal" or "warning signal," "aberrant signal," or "anomaly" in risk prevention. The article by Igor Ansoff, "Managing Strategic Surprise by Response to Weak Signals" [3], remains the reference in the research on weak signals. The concept of weak signals is identified in Management Sciences for the first time. Ansoff defines weak signals as *the early warning symptoms of strategic discontinuities, which act as low-intensity information that can indicate a trend or a significant event*. His article focuses on the need for a company to find weak signals to avoid threats or, conversely, to leverage opportunities. The publication of this article takes place after the first oil shock of 1973 when political instability demonstrated that the strategic plans established during the "Thirty Glorious Years" were no longer valid. Companies can no longer rely solely on extrapolating from past data. To avoid being caught off guard by environmental changes, they must anticipate and deal with the growing turbulence of the environment. Ansoff proposes a gradual response to detected threats or opportunities based on the amplification of weak signals. Ansoff's definition is based on the relevance of a weak signal by identifying it as an element with anticipatory characteristics. However, this definition is not sufficiently precise; it is more of a metaphor. Subsequently, many authors have built upon this work to refine their idea further.

Over the past fifty years, the definition of a weak signal has evolved. Before 1980, the concept of a weak signal referred to emerging phenomena that would have an impact in the future. In the 1980s, definitions focused on poorly defined sources and their impacts. During the 1990s, new adjectives emerged to describe why these signals are so tricky to detect: small, dynamic, and peripheral. From the 2000s onward, definitions began to refer to indicators of a phenomenon (such as a trend) rather than the phenomena themselves. On the following page, we present a chronology of key definitions. This list is not intended to be exhaustive; it is presented here merely to demonstrate the diversity of the concept of the weak signal. More recently, Van Veen and Ort [50] analyzed 152 articles, of which 68 defined the concept of weak signals. They aimed for a unified definition based on this study: *"A perception of strategic phenomena detected in the environment or created during interpretation that are distant to the perceiver's frame of reference"*. According to the following definitions, weak signals are precursors to an event and occur before a strong signal. A strong signal can be easily identified because its impact is clear and constant. The strong signal allows for a precise assessment of the situation and is seen as tangible evidence of a change [36].

As we have seen from the definitions discussed earlier, there are specific characteristics of weak signals that we need to consider to make this concept operational and thus allow its exploration. According to Hiltunen, weak signals can indicate many gradual changes. In 2008, she proposed to represent the weak signal in a three-dimensional space [21] where: 1) signal corresponds to an indicator

1990	Ansoff and Mc Dowell [4]	Imprecise early indications about impending impactful events.
1994	Godet [19]	A factor of change, hardly perceptible at present but which will constitute a strong trend in the future.
1997	Coffman [13]	An idea or trend that will affect how we do business, what business we do, and the environment in which we will work.
2003	Blanco et Lesca [6]	[. . .] weak signals announcing future major changes and to specify forthcoming events, their potential impact, importance and urgency.
2007	Macrae [33]	These doubts were produced through patterns of sense-making that interrelated organizational incidents with broader frames of reference in ways that made weak signals meaningful, relevant and worthy of further attention.
2008	Hiltunen [21]	Current oddities and strange issues that are thought to be key in anticipating futures changes in different environments.
2010	Kuosa [29]	Any observation which is totally surprising, amusing, ridiculous, or annoying to you.
2010	Hiltunen [22]	They are today's clues and signs that provide us with hints of possible future events and trends, page 74.
2011	Saritas and Smith [44]	Early signs of possible but not confirmed changes that may be later become more significant indicators of critical forces.
2011	Lesca et al. [31]	Outliers that are likely to constitute weak signals. The interpretation of outliers can highlight anticipative warning, surprises, inflexions, ruptures, fractures, discontinuities.
2012	Mendoça et al. [36]	Weak signals refer to premature, incomplete, unstructured, and fragmented informational raw material, that can be scrutinised, compiled, analysed and converted into an indicator of potential change.

of visibility, 2) event represents an indicator of the diffusion of a future sign, and 3) interpretation represents the understanding of the future sign by its receiver (see Fig. 1). Signs can be classified from weak to strong in her three-dimensional space: a weak signal has low levels for the signal and event dimensions and can transform into a strong signal with higher levels for these two dimensions. José Poças Rascão [40] indicates that signal strength is essential because signals range from weak to strong, eventually becoming a trend. This model has been a general framework in many works presented in Sect. 3. The signal dimension is used

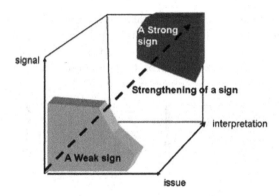

Fig. 1. Three-dimensional model proposed by Hiltunen, extract from [21].

to measure weak signals' visibility, which has a low frequency of occurrence, and the event dimension is used to quantify their diffusion. Diffusion is measured by the growth rate of the signal, which should increase for a weak signal. The three-dimensional model proposed by Hiltunen represents a first step towards an operational description of weak signals, which is necessary for an automated detection method. Based on the various definitions of weak signals in the literature and Hiltunen's representation, we have retained four characteristics of weak signals: **visibility**, **diffusion** (or emergence), **amplification**, and **rarity**. Accordingly, we propose our definition of a weak signal:

A weak signal is a signal that is barely visible and rare, indicating gradual changes, i.e., with increasing diffusion, that can lead to an event. McMinn et al. [35] define an event as an essential fact happening at a specific time and place.

In the context of Online Social Networks, events such as natural disasters, terrorist attacks, or the release of a new product brand trigger an unusual volume of messages. We apply the retained characteristics in Online Social Networks where interactions between individuals carry the signal. We then describe a weak signal by a small number of evolving interactions between individuals. Interactions between individuals are seen as particular motifs.

3   Weak Signals Identification Techniques

Identifying weak signals is a significant issue that has led to many research efforts, and various approaches and applications have been proposed. A comprehensive method for identifying weak signals consists of three steps, as shown in Fig. 2. The detection steps allow passing the surveillance filter corresponding to the ability of the signal to be detected. It is followed by the interpretation step that helps to pass the mentality filter, which is the ability of the signal to be recognized after being seen. Several cognitive biases, like normality or confirmation bias, were proposed to explain why weak signals are ignored [45]. The two steps leave the final decision lying in the decision-makers' hands, allowing it to pass the power filter [24]. In the following, some methods stop at the detection step and do not assist business experts in interpreting the detected weak signals.

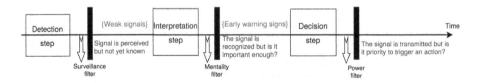

Fig. 2. Steps for identifying weak signals.

The vast majority of weak signal identification methods are based on document exploration through textual analysis of keywords, as two state-of-the-art reviews have pointed out in [37,43]. This involves working with the text by tokenizing it into words, removing stop words, and stemming. Many studies have also chosen to group keywords into topics to facilitate the interpretation of results. We have classified the existing methods into four families. Two large families of techniques were used to group keywords into topics: 1) Automatic Language Processing techniques and 2) Machine Learning techniques. The third family focuses on interactions between keywords and uses tools of graph theory (centrality, clustering coefficient) to detect weak signals. Finally, the last family of methods is specific to Economic Intelligence. They use impact analysis carried out by a Bayesian network to identify among documents weak signals and their consequences on the company. Figure 3 summarises the process of detection and interpretation of weak signal keywords.

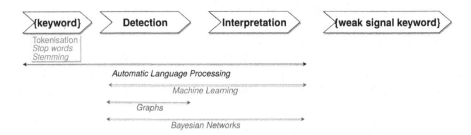

Fig. 3. Process of identifying weak signal keywords with the four families of techniques.

In Sub-sect. 3.1, we present techniques based on Automatic Language Processing because these techniques correspond to most works. In Sub-sect. 3.2, techniques based on graph theory are presented because these works are the closest to our BEAM method.

3.1 Techniques Based on Automatic Language Processing

Keywords are extracted from newspaper articles, research papers, or tweets. The TF-IDF (Term Frequency-Inverse Document Frequency) weighting method, which evaluates the importance of a word contained in a document, is generally used to select them. This method uses the following definition: *a keyword with a low visibility and a low degree of diffusion is considered as a weak signal.* To identify weak signals among them, most methods rely on the following two criteria:

1. the visibility of a keyword, evaluated by its frequency of appearance;
2. the diffusion of a keyword, evaluated by its growth rate seen as its evolution from one period to another.

Yoon in [53] formalized the visibility criteria of keywords appearing in documents and their diffusion over time with two metrics: 1) the degree of visibility (DoV) represents the frequency of the keywords in the set of documents; 2) the degree of diffusion (DoD) represents the frequency of documents in which a keyword appears. The two measures are formally defined for a keyword i and a period j, as follows:

$$DoV_{ij} = \left(\frac{TF_{ij}}{NN_j}\right) \times (1 - tw \times (n-j)) \qquad DoD_{ij} = \left(\frac{DF_{ij}}{NN_j}\right) \times (1 - tw \times (n-j))$$

with

- TF_{ij}, keyword i frequency during period j;
- DF_{ij}, frequency of the document in which i appears during the period j;
- NN_j, number of documents during the period j with n the number of periods;
- and finally, tw represents a weighting to take the time into account.

Yoon considers that recent occurrences of a keyword are more interesting than older keywords and sets tw to 0.05. A high and increasing frequency growth rate reflects the diffusion of the keyword/document. The average growth rate is calculated with the geometric mean given by $\mathring{X} = \sqrt[n]{\Pi_{j=1}^n x_j}$.

Yoon also proposed a first step toward the interpretation of weak signals. The idea is to provide a mapping of keywords, according to the two previous criteria, to business experts to help them interpret weak signals. The DoV and DoD measures are used to build, respectively, the KEM map (Keyword Emergence Map) and the KIM map (Keyword Issue Map), which are BCG matrices[3].

---

[3] The BCG matrix (Boston Consulting Group) is a strategic analysis tool from the end of the 1960s, it represents the way an organization develops its economic model https://fr.wikipedia.org/wiki/Matrice_BCG.

Figure 4 represents a BCG matrix divided into four zones containing the keywords. In the Weak signal zone, keywords (respectively documents) have a low frequency but a high growth rate, which may suggest that they will expand rapidly. In the Strong Signal or Trend zone, keywords (respectively documents) have a high frequency and growth rate. The third zone, Latent Signal, is characterized by keywords (respectively documents) with a low frequency and a low growth rate; the fourth zone, Well-known Signal, contains keywords (respectively documents) with a high frequency but a low growth rate.

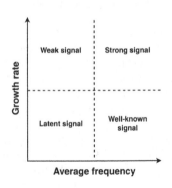

Fig. 4. Weak signals classification using a BGC matrix visualization.

First, this interpretation technique has drawbacks because some keywords were found on the borders of two zones. In addition, weak signals are isolated keywords that have lost their context, i.e., the documents in which they were found. A grouping of keywords into topics has been proposed to improve the interpretation of weak signals.

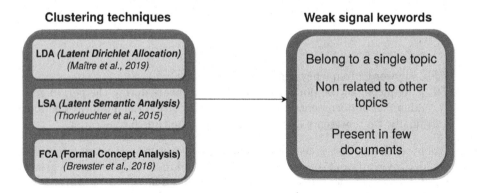

Fig. 5. Techniques that proposed keyword clustering into topics.

The left part in Fig. 5 shows three families of proposed techniques for clustering keywords into topics, and the right part is the definition of a weak signal keyword. LDA or Latent Dirichlet Allocation aims to infer the topics of a corpus of documents. In [34], Maitre et al. proposed using LDA combined with Word2vec[4] to detect a cluster or a topic related to a weak signal. LSA, or Latent Semantic Analysis, provides a low-dimensional representation of documents and words. It relies on a matrix that describes the occurrence of the word in the document

---

[4] Word2vec is a family of word embedding models creating vector representations of words, it is then possible to define a similarity value between two words.

(raw number or normalized by TF-IDF) and then uses the singular value decomposition (SVD) to reduce the number of words while preserving their relations. The method proposed by Thorleuchter et al. is based on the formulation of a hypothesis by business experts, this hypothesis being the occurrence of an event in the future. They use SVD to create word clusters and identify textual patterns corresponding to weak signals [49]. Brewster et al. [8] introduce an early warning approach for predicting organized crime as part of the ePOLICE project[5] They use FCA or Formal Concept Analysis to organize keywords in tweets based on selected properties such as temporal and/or geographic proximity between the same criminal activities. If the tweets containing these keywords are issued in the same geographical area and in a short period, they will be considered weak signals. When clustered, keywords are found in a single topic and are not linked to other topics.

3.2   Techniques Based on Graph Theory

Other methods have used basic tools of graph theory; their particularity is that they leave the interpretation task up to the business experts.

A graph is a set of vertices or nodes, connected or not by edges. Formally, a graph $G = (V, E)$ is represented as follows:

1. $V = \{v_1, v_2, \ldots, v_n\}$ the set of vertices/nodes representing entities or individuals that interact with each other;
2. $E = \{e_1, e_2, \ldots, e_m\}$ the set of edges representing a connection or interaction between the nodes, $e_i = (v_i, v_j) \in E$ is an edge that connects nodes v_i and v_j.

$|V|$ is the total number of vertices called order, $|E|$ is the total number of edges called size. The degree of a vertex v_i denoted $d(v_i)$ is equal to the number of edges that connect it to the other vertices; it is the number of its neighbors: $d(v_i) = |(v_i, v_j)|, (v_i, v_j) \in E, \forall v_j \in V$, Δ is the maximum degree of the graph. Betweenness centrality and closeness centrality are measures of centrality in a graph based on shortest paths. Betweenness centrality of a node v_k is given by:

$$C_B(v_k) = \sum_{v_i \neq v_k \neq v_j} \frac{\sigma_{v_i v_j}(v_k)}{\sigma_{v_i v_j}},$$

where $\sigma_{v_i v_j}$ is the total number of shortest paths from node v_i to node v_j and $\sigma_{v_i v_j}(v_k)$ is the number of those paths that pass through v_k (not where v_k is an end point).

Closeness centrality of a node v_k is given by:

$$C_C(v_k) = \frac{1}{\sum_{v_i} d(v_i, v_k)},$$

---

[5] early Pursuit against Organized crime using envirOnmental scanning, the Law and IntelligenCE systems https://cordis.europa.eu/project/id/312651.

where $d(v_i, v_k)$ is the distance, i.e., the length of the shortest path between vertices v_k and v_i.

Dotsika et al. [15] look to identify indicator keywords for potentially disruptive technologies. A co-occurrence graph is created, and three classical centrality measures are used: degree centrality, betweenness centrality, and closeness centrality. These measures correspond to the popularity of the keywords/topics and their influence on the trends. Table 1 interprets the measures two-by-two w.r.t the notion of trend. The red-colored cells correspond to keywords that are:

- infrequent, i.e., low degree, but form important topics, i.e., high closeness degree;
- keywords that link a lot of topics, i.e., high betweenness degree, and form a nest, i.e., low closeness degree.

Keywords in these colored cells are considered to announce trends, i.e., weak signals potentially.

Table 1. Detecting weak signals with centrality measures

High ＼ Low	Degree	Betweenness degree	Closeness degree
Degree		Popular and frequent keyword where links bypass it	Popular keyword seen as a niche, integrated in a cluster thematically distant from the rest of the graph
Betweenness degree	Infrequent keyword with few links but important for flows		Keyword seen as a nest and monopolizing links between strong and marginal trends
Closeness degree	Infrequent keyword linking main topics	Central and known keyword appearing on many paths	

Kwon et al. [30] use betweenness centrality and the Minimum Spanning Trees (MSTs)[6] to detect weak signals. The authors build a matrix from the co-occurrences between keywords corresponding to products and services provided by companies. This matrix is seen as a graph where the betweenness centrality of each node is computed. Consequently, a weak signal appears as the node having the smallest value of betweenness centrality. The graph being large, MST offers a more compact visualization of the graph.

After examining the existing weak signal detection techniques, we found that it is difficult to apply them on data coming from Online Social Networks. Indeed, most of the methods rely on keywords or keyword clustering, but the data coming from OSNs are composed of short texts (for example 280 characters maximum in a tweet), and often include abbreviations and spelling errors. In addition, social

---

[6] Given a connected undirected graph whose edges are weighted, the Minimum Spanning Tree of this graph is a tree that connects all vertices, whose sum of the weights of the edges is minimal, i.e., of weight less than or equal to that of all other spanning trees of the graph.

relations are based on interaction links between entities, therefore we should rely on graph theory techniques. However, existing techniques still use already-known tools, such as the centrality measures presented above. Moreover, the computation of these measures on large graphs is often very expensive.

A graph of temporal interactions often models an Online Social Network. This representation leads us to hypothesize that we can explore elements from the network topology to identify weak signals. Works using graph theory [5,14,26] also endorsed our hypothesis that graphlets, particular network motifs, can be precursors of events. We also want to propose tools for interpreting weak signals beyond keywords clustering so that the notion of weak signals can be used as a decision-support tool. In the following, we introduce our method BEAM, which detects and interprets weak signals in a network of interactions.

4 BEAM: A Framework for the Detection and Interpretation of Weak Signals

We propose a framework, BEAM, that acts as a systematic process for detecting and interpreting weak signals using data modeled as a graph of temporal interactions. Data from Online Social Networks like Twitter or Facebook, telecommunication networks, or financial transactions can be seen as temporal graphs. Each interaction takes the form of a link between two entities appearing at the time of the interaction. BEAM implements a quantitative method to automate the process of detecting weak signals and uses qualitative tools offering complementary elements to business experts to help them make decisions.

BEAM is based on a standard data processing pipeline that includes the collection and exploration of data before weak signal detection, as outlined below. In the first step, we process the collected raw data and build the study corpus, which is then divided into snapshots of the same duration for a more detailed study.

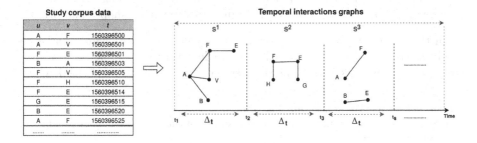

Fig. 6. Creation of snapshots with same duration Δt, from the study corpus.

BEAM takes as input a list of temporal interactions where each row represents a three-component tuple of the form (u, v, t), such as (u, v, t) indicates that

u interacts with v at t where $t \in [\alpha, \beta]$, and α and β correspond to the timestamps in the studied period. As our method is based on the network topology and time is a significant element, we have chosen to represent the temporal interactions as a sequence of $s \in \mathbb{N}$ static graphs $\mathfrak{G} = \{S^i | i \in \{1, \ldots, s\}\}$ where S^i, denoted snapshot i, is the undirected and unweighted graph containing all interactions that occurred between the times $t_i = \alpha + i\Delta t$ and $t_{i+1} = \alpha + (i+1)\Delta t$. Δt is a constant duration for all snapshots representing one day, one hour, thirty minutes, ten minutes, etc. The goal of Δt is then to connect nodes as a function of time, such that two consecutive snapshots S^i and S^{i+1} are Δt-adjacent. Figure 6 presents a list of tuples (u, v, t) ordered by t^7, which is decomposed (on the right of the figure) into a series of snapshots of same duration Δt. In this figure, snapshots have different sizes, and Δt is equal to 6.

In the second step, we explore the data of the study corpus by applying algorithms on each snapshot to analyze the network topology. Indeed, we hypothesize that the network topology plays a vital role in information propagation. We choose **graphlets, which are small patterns, as an operational description to detect weak signals** in a temporal interactions graph.

Graphlets are induced subgraphs[8] connected and non-isomorphic[9], ranging from 2 to 5 nodes and chosen among the nodes of a large graph. They were first introduced in 2004 in the field of molecular biology [39]. There are thirty different types of graphlets labeled G_0 to G_{29}[10]. An essential element in the notion of graphlets is the orbits. Orbits, labeled O_0 to O_{72}, are the equivalence classes of graphlets and correspond to the position or role of nodes in graphlets (see Fig. 7). In the same graphlet, nodes belonging to the same orbit (same color) are interchangeable; in other words, they can play the same role in the network. Graphlets are good summaries of the variability of interactions in social networks [46], and they have characteristics generally associated with weak signals. They are:

- fragmentary and not very visible because they are small sub-graphs;
- not very significant, taken in isolation, a sub-graph of at most five nodes does not mean much in the voluminous data produced by the social network;
- explainable by business experts through their predefined shapes and orbits.

In the third step, we aim to establish a signature of weak signals based on graphlet features. Next, in the fourth step, we define criteria to qualify the identified precursors as weak signals. The final step is the interpretation of the detected

---

[7] t component has the timestamp epoch format.

[8] In graph theory, an induced subgraph is a subset of the nodes and **all** their edges in the original graph.

[9] In graph theory, an isomorphism of two graphs, G and H, is a correspondence between the sets of nodes in G and H, such that if two nodes are adjacent in G, they are adjacent in H. Graphlets are non-isomorphic because they do not have the same shape.

[10] In this document, we use the term graphlet for each type among the 30, however, this does not represent its occurrence.

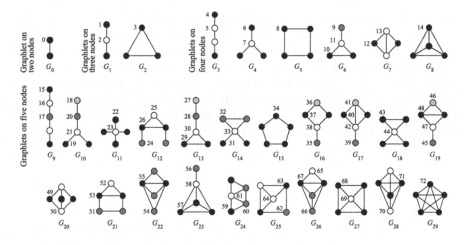

Fig. 7. 30 graphlets and 73 orbits.

weak signals. Graphlets leave room for interpretation by business experts, thus eliminating the black box effect that an entirely automated method might have.

The steps mentioned above can be iterated at different levels: 1) steps two to five are iterated on each snapshot until the weak signals become strong signals or false alarms; 2) the detection of weak signals and their interpretation can call into question the study corpus and the process then starts again at the first step with the modification of the study corpus.

5   Use Case: Game of Thrones' Final Season

With high-quality streaming, HBO[11] estimated that the final season of the television series Game of Thrones (GOT) averaged 44.2 million viewers per episode. The final season resulted in much chatter due to writer changes and the shortening of the season, with only six episodes airing once a week between April 14, 2019, and May 19, 2019. The episodes aired live at 9 p.m. U.S. time and the collected data correspond to tweets posted at the same timestamp.

The aim of the study conducted on this data set is to examine the ability of BEAM to detect weak signals before a significant and viral event presented in Sub-sects. 5.2 and 5.3.

5.1   Data Preparation

The GOT raw data set consists of tweets published between April 10, 2019, and May 25, 2019, before and after the launching of the final season. The criteria used to collect this data are hashtags #gameofthrones, #got and #gots8 (for the eighth season of the series) to eliminate tweets unrelated to the series. The

---

[11] Home Box Office is a U.S. pay-TV network owned by WarnerMedia Studios.

number of collected tweets is equal to 46,481,705 issued by 8,194,319 users. We choose to work on the retweets relation because it represents information diffusion and reflects the virality of this event. In addition, only retweets posted on the day each episode aired between 12 pm and 8 pm are retained (i.e., one hour before the episode airing). This phase is essential because identifying weak signals makes sense in a well-defined context only, and the quality of the filtered data determines the reliability of the decision-making process. Finally, each retweet collected in JSON format is transformed into a tuple in the form (u, v, t) as shown in Fig. 8.

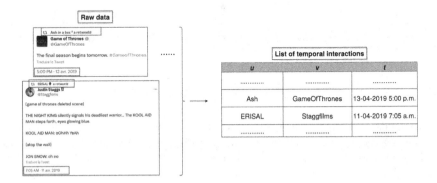

Fig. 8. Modeling the raw data of tweets into a list of temporal interactions.

We created six study corpora, one per episode at the end of this process. Table 2 shows the size of six study corpora in terms of retweets' number and size of graphs.

Table 2. Properties of the graphs created based on the retweets in each episode.

Episode	Study period	Retweets	Nodes	Links
1	April 14 12 p.m. → 8 p.m.	248,919	200,390	237,048
2	April 21 12 p.m. → 8 p.m.	75,157	67,408	71,977
3	April 28 12 p.m. → 8 p.m.	154,423	129,112	148,856
4	May 5 12 p.m. → 8 p.m.	79,686	68,357	76,111
5	May 12 12 p.m. → 8 p.m.	56,628	53,535	54,239
6	May 19 12 p.m. → 8 p.m.	253,604	201,909	242,841

These corpora are then decomposed into snapshots of the same duration Δt equal to 10 min. The choice of the value of Δt is a trade-off between three factors: 1) the volume of data in the corpus, 2) the density of the links in the sub-graphs of the snapshots and 3) the time needed by the analysts to interpret the detected weak signals. Indeed, small snapshots may ignore weak signals, while large snapshots or a dense sub-graph will not grant the analysts enough time to interpret the signals and raise the alert. On the other hand, if the sub-graph has a low density, then it could omit complex graphlets that might provide many insights during the interpretation stage. As the episodes are broadcasted around 9 p.m. once per week, for each episode, we applied BEAM on thirty-five snapshots having the same duration between noon and 8 p.m. To justify our choice of snapshots duration, we experimented on the first episode's study corpus while varying the duration of the snapshot between 10, 15, 30, 40, 50, and 60 min. As per the obtained results, we choose the duration of the 10-minute snapshot since it allows us to find the most significant number of weak signals carrying the information in the volume of data.

5.2   Detection of Weak Signals

In this section, we present a description of weak signals via their visibility, diffusion, amplification, and rarity. Then as a use case, we will describe the results obtained on the study corpus corresponding to April 14th, the release date of the first episode, and for the snapshot from 5:00 pm to 5:10 pm, which contains 7,898 nodes and 7,109 links.

Signal Visibility. First, the signal visibility criterion is estimated by the number of graphlets. A state-of-the-art on graphlet enumeration was carried out by Ribeiro et al. in 2019 [41], in which they presented an overview of existing algorithms and highlighted their main advantages and limitations. To choose the most appropriate algorithm to enumerate graphlets and orbits in the studied snapshots, we defined four criteria:

1. the exact enumeration of graphlets up to five nodes;
2. the enumeration of orbits;
3. an acceptable complexity;
4. the availability of the source code.

The first criterion ensures the completeness of the graphlets. The second criterion provides the interpretability of the results by studying the shape and the role of the nodes in the graphlets. The two other criteria guarantee that the proposed method is materializable. We, therefore, choose Orca algorithm [23] since it meets the four required criteria. Its complexity is $O(|E| \times \Delta^2)$ and its source code is available on the following link: https://rdrr.io/github/alan-turing-institute/network-comparison/src/R/orca_interface.R.

The thirty graphlet types are enumerated with Orca for each snapshot S^t of the study corpus. As a result, each snapshot is represented as thirty components of a numerical vector $(G_0^t, G_1^t, ..., G_{29}^t)$, where G_x^t is the number of graphlets of type x in snapshot S^t.

We then apply a normalization procedure on the calculated number of graphlets to re-scale their values to be proportional to each other. Even if the snapshots have the same duration, the number of nodes and corresponding links vary from one snapshot to another. For example, some snapshots have only a few links, while others have thousands. This step is significant because it should not mask the weak signals but make them comparable. The chosen procedure is the one proposed by D. Goldin and P. Kanellakis, in which they study the similarity between two queries on a temporal database [20]. By applying this normalization procedure for each of the snapshots S^t where s is the number of snapshots, each component G_x^t of its vector with $x \in \{0, \ldots, 29\}$, is normalized by:

$$\overline{G_x^t} = \frac{G_x^t - \mu(G_x)}{\sigma(G_x)}$$

with $\mu(G_x)$ the mean of each graphlet G_x for all snapshots, given by:

$$\mu(G_x) = \frac{1}{s} \sum_{t=1}^{s} G_x^t$$

and $\sigma(G_x)$, the standard deviation given by:

$$\sigma(G_x) = \sqrt{\frac{\sum_{t=1}^{s}(G_x^t - \mu(G_x))^2}{s-1}}$$

Signal Reinforcement: Diffusion and Amplification. The next step is the estimation of the signal reinforcement. The velocity and acceleration evolution are quantitative features that allow to evaluate the signal's diffusion and amplification. We use them to identify event precursors among the graphlets. We define an event precursor as an observable and clear fact that exists in an organization's business process and is caused by existing factors in the process [28]. From the normalized values, the evolution of all the G_x^t components is studied via their velocity of appearance and their acceleration. Our goal is to highlight the graphlets that emerge before the others. For each snapshot and each graphlet G_x, we derive its velocity as the difference between the normalized value of the graphlet at snapshot S^{t+1} and the normalized value of this same graphlet at snapshot S^t:

$$\overline{V_x^t} = \overline{G_x^{t+1}} - \overline{G_x^t} \quad \forall x \in \{0, \ldots 29\}$$

The acceleration is computed in the same manner by calculating the difference between velocities:

$$\overline{A_x^t} = \frac{\Delta V_x}{t} = \overline{V_x^{t+1}} - \overline{V_x^t} \quad \forall x \in \{0, \ldots 29\}, t = 1 \text{ snapshots are of same duration.}$$

The output is thus a numerical matrix representing for each snapshot S^t, the normalized value of the graphlets $\overline{G_x^t}$, their velocity $\overline{V_x^t}$, and their acceleration

$\overline{A_x^t}$. The velocities can only be computed from the second snapshot and the accelerations from the third.

At the end of this phase, we identify a list of precursors among the graphlets. Those emerge and diffuse more rapidly than the others, which is determined according to the values of velocities and accelerations. To do this, we propose two options. The first option consists of the following for the studied snapshot S^t:

- rank their velocities $\overline{V_0^t}, \overline{V_1^t}, \overline{V_2^t}, \ldots, \overline{V_{29}^t}$ in descending order;
- fix a top k value for which the k graphlets have the highest velocities. Similarly, we set a top k for graphlet accelerations.

For El Akrouchi et al. [17], weak signals do not represent more than 20% of the information, which complies with the Pareto principle, setting the value of k to 6 in BEAM. The second possibility consists of setting a threshold in the studied snapshot, equal to the average of the velocities, for example, and selecting those greater than or equal to this threshold, $\overline{V_x^t} \geq \mu(\overline{V^t})$, a similar threshold is set for the accelerations.

We then suggest a visualization tool to help business experts identify precursors. We rely on Hiltunen's approach [21] in which she considers two types of precursors: **early information** and **first symptom**. The first type represents sudden new information, such as announcing a new product or invention. In contrast, the second type represents a remarkable change that is difficult to interpret. The visualization tool is designed as a BCG matrix divided into four zones (not necessarily the same size) according to the value chosen for the top k or the threshold.

Figure 9 illustrates the BCG matrix created for the 5:00 to 5:10 p.m. snapshot of April 14th. It represents on the horizontal axis the diffusion of the signal measured by the velocity of graphlets, and on the vertical axis, its amplification measured by their acceleration. The matrix is divided here according to thresholds: respectively equal to 0.02 for the velocity and 0 for the acceleration. We find the two types of precursors in the matrix's blue-shaded zones that Hiltunen considers. These zones contain graphlets with 1) high velocity and low acceleration, 2) low velocity and high acceleration, 3) high values for both criteria. The graphlets in these three zones are selected as precursors of events because they are observable facts and remarkable by their velocities and/or accelerations. The area not shaded in blue is considered noise, and contains graphlets with very small or negligible velocities and accelerations. Noise represents unclear information that appears randomly but does not make sense. Mendoça et al. [36] see noise as irrelevant signals pointing in inconsistent directions.

Signal Rarity. To qualify graphlets as weak signals, we rely on the rarity criterion. This criterion is estimated by a ratio measuring the proportion or contribution of each of the thirty graphlets to the total number of graphlets. We propose a global ratio calculation in which the total number of a graphlet G_x in the studied s snapshots is divided by the total number of all graphlets for these

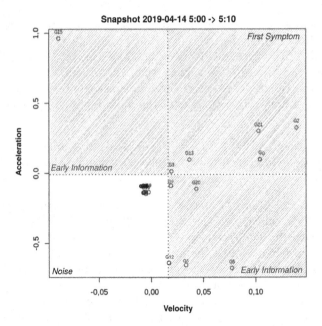

Fig. 9. BCG matrix of graphlets' velocities and accelerations in the snapshot from 5:00 pm to 5:10 pm

s snapshots:

$$R_{Global}(G_x) = \frac{\sum_{t=1}^{s} G_x^t}{T_{Global}(G)} \ \forall x \in \{0, \ldots, 29\}$$

where $T_{Global}(G) = \sum_{t=1}^{s}(\sum_{x=0}^{29} G_x^t)$, s is the number of studied snapshots.

As weak signals are rare, they are the ones that participate the least, so the resulting ratio is sorted in ascending order. This ratio calculation allows a classification of the graphlets in four categories listed in Table 3. True Positives are precursor graphlets with a low ratio. False Positives are graphlets identified as precursors but are not weak signals and must be discarded at the end of this step. True Negatives are graphlets that are neither precursors nor weak signals, and False Negatives are not identified as precursors but have a low ratio. This classification preserves True Positives and introduces False Negatives as weak signals. Retaining some False Positives could help business experts explore other latent information.

Table 3. Classification of graphlets in four categories.

	Weak signal	¬ Weak signal (noise or strong signal)
Precursor	True positive	False positive
¬ **Precursor**	False negative	True negative

The first five weak signal graphlets detected in this snapshot are listed in Table 4 and highlighted in red, with their shape, number, velocity, acceleration, and ratio. Referring back to the matrix in Fig. 9, we note that graphlets G_{12} and G_9 are at the border of the noise and early information zones, with low velocity and acceleration values. However, the contribution ratio calculation used to estimate the rarity criterion, confirmed them as weak signals and thus classified them as true positives. On the other hand, graphlets G_3 and G_{15} only have a high value in one of the criteria (either high velocity or acceleration) but a high ratio. According to the rarity criterion calculation, these two graphlets are discarded and classified as false positives.

Table 4. Five weak signal graphlets (highlighted in red) detected in the snapshot of 5:00 pm to 5:10 pm, as well as two other graphlets discarded by BEAM.

Graphlet	G_{12}	G_6	G_9	G_{21}	G_{13}	G_3	G_{15}
Shape							
Number$_{(Visibility)}$	959	112	1 917 278	17	402	146 773	33
Velocity$_{(Diffusion)}$	0.001	0.009	0.005	0.097	0.035	0.018	-0.086
Acceleration$_{(Amplification)}$	0.004	0.003	0.004	0.285	0.091	0.012	0.932
Contribution$_{(Rarity)}$	0.052	0.066	0.076	0.081	0.089	0.092	0.094

Confirmation of Weak Signals. Once weak signals are detected, experts have to provide proof of their existence and their role in the changes that have occurred in the environment. This requires the occurrence of an event to justify the intrinsic properties of BEAM by confirming a dependency between the graphlets and an event in the study corpus and by demonstrating this dependency's eventual time shift. An event is considered a situation where the number of interactions reaches its maximum value. We, therefore, propose a statistical method, the cross-correlation [12], associated with a visualization tool.

In this step, we model the data as a time series. On one hand, the series of the corpus data is a sequence of n interactions representing the retweets that occurred over the study period: $X = (x_t)_{1 \leq t \leq n} = (x_1, x_2, \ldots, x_n)$, where x_t is the number of retweets in snapshot S^t. On the other hand, graphlet data is modeled as time series, representing sequences of n values corresponding to the number of each graphlet type in each snapshot (left part of Fig. 10). We must have a dependency/correlation with a negative offset to confirm the graphlets as weak signals. The event is considered to be the diffusion of the first GOT episode, which occurred at 9:00 pm. The correlation coefficient is between -1 and $+1$, quantifying how the two series vary. A high positive value indicates that the two series vary enormously, a low positive value means that they vary together but with a deviation. In contrast, a high negative value indicates that they vary in opposite directions but still linearly. We consider having n observations in the retweets series X and that of a graphlet G_x, $Y = (y_1, y_2, \ldots, y_n)$. The cross-correlation coefficient ρ of the two series X and Y, at a time offset l is defined with:

$$\rho_{xy}(l) = \frac{\sum_{t=1}^{n}(x_t - \overline{x}) \times (y_{t-l} - \overline{y})}{\sqrt{\sum_{t=1}^{n}(x_t - \overline{x})^2} \times \sqrt{\sum_{t=1}^{n}(y_{t-l} - \overline{y})^2}} \tag{1}$$

In this equation, x_t and y_{t-l} are the observations of the two series, where the second is shifted of l w.r.t the first one. $\overline{x}$ and $\overline{y}$ represent the averages calculated for each of the series. The numerator corresponds to the covariance between X and Y, the denominator corresponds to the product between the standard deviations of the two series, being always greater than the numerator, the value of the correlation will not exceed 1. The covariance shows whether two variables tend to move in the same direction, while the correlation coefficient measures the strength of this relationship on a normalized scale, between -1 and $+1$ [12]. To visualize correlation matrices, correlograms are used, where a correlogram is a graphical representation of the correlation coefficient at each offset [18]. This representation highlights the maximum value of cross-correlation, being at the offset where the time series are the most similar.

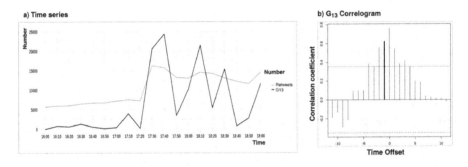

Fig. 10. Correlation of G_{13} graphlet with the retweets time series.

We observed the created correlograms in the studied snapshots. The right part of Fig. 10 shows the correlogram of G_{13} weak signal graphlet, presenting a positive correlation with a negative offset. This correlogram represents in the x-axis the time offset and the correlation coefficient in the y-axis. The value of the coefficient at an offset equal to 0 is the reference point of the correlogram because it represents the event. The red line shows a positive correlation of 0.63 with a negative offset $= -1$. This correlation confirms the G13 graphlet as a weak signal.

In this subsection, we have proposed graphlets as an operational tool to establish a signature of the weak signal, which allows quantifying its four characteristics:

low visibility, diffusion and amplification over time, and rarity. Figure 11 summarizes the detection steps described.

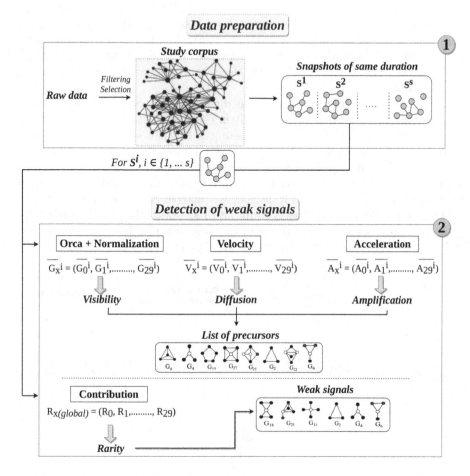

Fig. 11. Summary of the weak signals detection steps in our method BEAM.

The following section presents the tools we provide business experts to interpret the detected weak signals.

5.3 Interpretation of Weak Signals

This step aims to help business experts decide if the weak signal can constitute an early warning sign. We define an early warning sign as a weak signal whose interpretation by business experts indicates that an event may be occurring which could be of great importance to the organization. Here the role of the business experts is decisive: they must determine the relevance of the relationship between the weak signal and a threat or an opportunity for the organization. This step is not entirely automatic, requiring business experts, who should be able to decide at the end if the weak signals should be transmitted to the decision-makers or ignored. The detected weak signals in the previous section constitute fragmented information and are therefore difficult to interpret. However, combining these signals with contextual elements or insights will give them meaning and relevance to enlighten the decision-making of business experts and decision-makers.

In this step, we propose analysis on different levels of granularity. On the graphlet instance or occurrence level, it is possible to study the relationships between the nodes belonging to the instances of weak signal graphlets using Cypher queries executed on a Neo4j graph database or SQL queries executed on a PostgreSQL relational database. On a finer level, the properties of the nodes, such as their type or activity on the social network, are studied. On the graph level corresponding to the study period, we analyze the communities and the ranking of the nodes using the Louvain and PageRank algorithms. Finally, we provide diagrams to visualize the data on these different levels of granularity. The provided analysis process and tools are illustrated in Fig. 12.

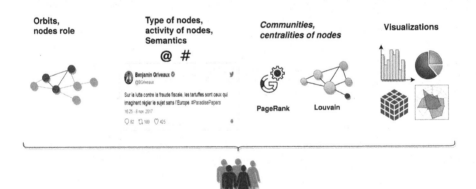

Fig. 12. Interpretation of weak signals using different levels of granularity.

From Types of Graphlets to Instances of Graphlets. We moved here from types of weak signal graphlets to instances of weak signal graphlets by studying the relation between nodes belonging to weak signals, as well as their role. The role of a node is characterized by its orbit or position in the graphlet. Since Orca enumerates the orbits of each node, we can identify different roles of nodes, such as central, intermediate, or peripheral. We will now present our analysis on the nodes belonging to the orbits of some weak signal graphlets. Listing 1.1 is an analytical query executed on the PostgreSQL relational database, returning user accounts that belong to the orbits of graphlets G_9 and G_{13} at 5:00 pm. The studied snapshot and the graphlet types are indicated in the WHERE clause, with the list of orbits to which the users belong.

```
SELECT u.user_screen_name,
       u.user_type,
       STRING_AGG(g.graphlet_type, ',', order by g.graphlet_type)
    as GRAPHLET_LIST,
       STRING_AGG(g.orbit_type, ',', order by g.orbit_type) as
    ORBIT_LIST
FROM User u, Graphlet g, Orbit o, Snapshot s
WHERE s.orbit_num = o.orbit_num
  AND s.graphlet_num = g.graphlet_num
  AND s.user_id = u.user_id
  AND o.graphlet_type = g.graphlet_type
  AND g.graphlet_type IN ('G9', 'G13')
  AND s.snapshot = '5:00'
GROUP BY u.user_screen_name, g.graphlet_type, g.orbit_type
ORDER BY u.user_screen_name;
```

Listing 1.1. SQL analytical query returning nodes appearing in weak signals orbits.

Table 5 is an extract from the result of the above SQL analytical query. It shows the properties of user accounts and the orbits of weak signal graphlets that they occupy in the studied snapshot.

Table 5. Extract of important nodes belonging to the orbits of detected weak signal graphlets.

User account	Account type	Graphlet	Orbit
@Got_Tyrion	Lord tyrion lannister (character)		O_{15}
@FreeFolkMemes	GOT memes account		
@TheMasters	Official account of golf championships	G_9	O_{16}
@jonatas_maia12	Product designer		
@GameofThrones	GOT official account		O_{17}
@TylerIAm	Journalist, Youtuber		
@JeffMillerTime	Marketing Manager at Snapshat	G_{13}	O_{27}
@9GAG	Online platform, viral and funny videos		

Another way of studying instances is to implement Cypher queries with a Neo4j database. Cypher queries allow us to visualize the relationships between nodes in graphlets and the participation of a node in all instances of the same graphlet type. Listing 1.2 is a Cypher query that returns twenty instances of the G_9 weak signal graphlet in the studied snapshot.

```
MATCH (u1)--(u2)--(u3)--(u4)--(u5)
WHERE NOT ((u1)--(u3)) AND NOT ((u1)--(u4)) AND NOT ((u1)--(u5))
    AND NOT ((u2)--(u4)) AND NOT ((u2)--(u5)) AND NOT ((u3)--(u5))
RETURN * LIMIT 20
```

Listing 1.2. Cypher query returning twenty instances of G9 graphlet.

Here, the business expert can get a global view of the membership of nodes in graphlet orbits (the same result as the above analytical query) but in the form of a graph. In addition, Neo4j offers the possibility of filtering a particular instance for a more detailed study by the expert. In the above Cypher query, which returns 20 instances of G_9, adding conditions, like node names, to produce a single instance is possible. Indeed, on the relational database level, we can study the frequency of appearance of a node in orbit, whereas, on the graph database, we can study the direct relationships of a node with other nodes in a graphlet.

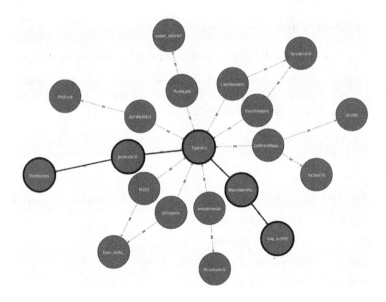

Fig. 13. Twenty instances of G_9 graphlet weak signal, of users connected with @TylerIAm, in the 5 p.m. snapshot.

Figure 13 shows the result returned by the above Cypher query in the form of a graph. In this figure, we find @TylerIAm in the central position, surrounded by fans, blogger accounts, or online platforms that might have retweeted him. This figure highlights one of the instances of G_9 in black. We notice that @Woodlawnwonder blogger and @joestudz18 artist and concept illustrator occupy the intermediate position (orbit O_{16}) in this graphlet. The peripheral position (orbit O_{15}) is occupied by the accounts @holy_schnitt social media star and @TheMasters major golf championship[12]. This instance is interesting since it highlights the activity of bloggers or stars who criticized the script and the actors' performance. Twitter users highly follow these accounts, and analysts should monitor their activity. This visualization also shows directly or indirectly connected users in weak signal graphlets.

Qualitative Analysis. This analysis lets business experts zoom in on moments of controversy and explore specific issues. In the following, we present a detailed analysis of the activity on some of the accounts mentioned above who played support or criticism roles towards the diffused information. We then translate this activity into a story of the tweets published by these accounts before the first episode's broadcast, which were retweeted quickly or became viral in the studied snapshot.

Two days before the episode aired, @9GAG issued a tweet (first tweet to the left of Fig. 14) in which he criticizes the set of GOT, which is similar to the set of the animated film Shrek. At the top of his tweet, he also compares the character Jamie Lannister from GOT with another character from Shrek. This similarity was also noticed by a mega-fan of GOT, who posted on Twitter, "So basically Game of Thrones is a Shrek live-action", with a figure showing side-by-side scenes of GOT and the movie Shrek, to express that GOT is an imitation of Shrek. His tweet quickly took off with over 47,000 retweets. On April 13, one day before episode 1 airing, @JeffMillerTime tweeted around 10:30 p.m. His tweet (second tweet in Fig. 14) contains a video in which we find Times Square in New York, then a dragon that appears in the sky, and the famous Flatiron building gets completely covered with ice. Jeff Miller took advantage of the viral nature of GOT to encourage viewers to use his augmented reality platform. This tweet has been retweeted 3,027 times until the end of the last season. A few hours later, @TylerIAm tweeted. Trill Withers is a journalist who was formerly a commentator on a sports channel, then a host of live shows on his YouTube channel, Monday to Friday from 12 p.m. to 3 p.m. His tweet (third tweet in Fig. 14) highlights the comic aspect of two main characters and announces unexpected changes. Retweeted about 17,000 times, we experimented with studying its virality. For this, we have calculated the duration from which it reaches 10%, 20%, up to 100% of its retweets. According to this experiment, 50% of the retweets are reached in about 20 min, and 24 h later, 95% of retweets are reached. These results highlight the role of Twitter in the fast propagation of information that can constitute a threat or an opportunity. @TheMasters also posted a tweet on April 13, one day before the last Golf Championship tournament and the broadcast of the first episode of GOT. This tweet went viral the next day as many golf and Game of Thrones fans retweeted it. Having won the tournament one step ahead of the three finalists, Tiger Woods's victory as an American professional golfer sparked excitement among spectators who tweeted about golf and the Game of Thrones series. The last two tweets at the

---

[12] https://en.wikipedia.org/wiki/Masters_Tournament.

right of Fig. 14 were issued two hours before the episode aired by fans of golf and Game of Thrones, who were eagerly waiting for both Tiger Woods to play and the start of the first episode of GOT.

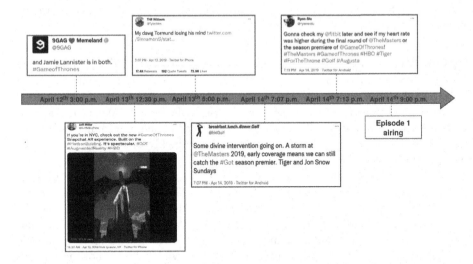

Fig. 14. Story of important accounts' tweets, prior to the airing of GOT's first episode.

As shown in this figure, we have classified the accounts posting the tweets into two categories. The weak signals carried by the accounts whose tweets are bordered in red can threaten the producers because they make fun of GOT or distort its image. Therefore, their activity must be monitored to mitigate their negative impact on the audience. On the other hand, those bordered in green are considered opportunities, as the accounts here are using GOT's reputation to advertise their platform or promote their topics of interest.

Macroscopic Study. We also propose a macroscopic approach to understand the broad context of the elements appearing in the weak signals: nodes' centrality to determine their influence and community to determine their position in the graph of the studied period.

We applied the Louvain algorithm [7] on the graph of the study corpus, which detected 1,800 communities, the largest of which contains about 9,000 nodes. It is also possible to study the ranking of nodes in the corpus of study or each community. We use the PageRank algorithm [9] to measure the centrality of nodes. This feature provides experts with information on the influence of particular nodes concerning others. Figure 15 shows some nodes belonging to the most prominent communities distinguished by different colors.

The yellow community regroups the accounts of GOT characters like @Daenerys and @Got_Tyrion. The green community represents accounts of fans in England as well as @skyatlantic, the English television channel. The pink

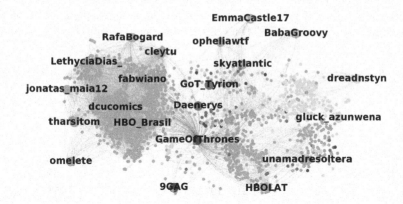

Fig. 15. Important communities detected by the Louvain algorithm. (Color figure online)

community includes accounts in Nigeria, where one of them, @dreadnstyn, posts and reacts to funny viral tweets. Among the nodes belonging to the orange community, we find the account of a blogger @unamadresoltera and the official site of HBO in Latin America. The largest community in blue contains nodes belonging to the weak signal graphlets, notably some found in the graphlet G_{13} at 5:00 pm. These nodes are accounts located in Brazil, including a comedian and digital creator @cleytu, a host @fabwiano, the Brazilian version of the HBO channel @HBO_Brasil, a journalist @LethyciaDias_, the biggest entertainment site in Brazil that broadcasts TV series, movies and music @omelete, @dcucomics the fan account of the DC universe (fictional universe produced by Warner Bros), @tharsitom a mega fan of books and novels, etc. We notice that all the communities are linked to each other through @GameofThrones, which is the official account of GOT. The communities show that the geographical location of the accounts is crucial in the communication, which the analysts must take into account.

6   Validation of BEAM

The purpose of this section is to validate our BEAM method. For this, we carried out two studies: one to show the results replication on periodic data that are the six episodes of Game of Thrones (Subsect. 6.1), and another one to verify the robustness of the method by varying the graph of interactions (Subsect. 6.2).

6.1   Studying the Results Replication

To verify the replication of the results, we performed the same experiment on retweets posted in the remaining five episodes of GOT. To avoid repeating the experimental steps described above, we explained how a few accounts carrying weak signals identified in the first episode evolved over the six episodes. Table 6 presents the number of retweets in each episode and the frequency of essential user accounts appearing in the orbits of the weak signals, episode by episode.

Table 6. Evolution of users appearing in weak signal orbits, episode by episode.

User	Episode 1	Episode 2	Episode 3	Episode 4	Episode 5	Episode 6
9GAG	4,497	0	48,584,856	923,971	184,262	0
Daenerys	879,983	0	32,778,147	0	0	4
FreeFolkMemes	26,880	240	8,072,963	0	0	0
GameOfThrones	3,825,049	0	0	1	25	0
HBO_Brasil	807,081	277,212	107,278,620	246	2,158	0
LordSnow	0	22,820	42,253,437	350,213	0	8
Retweets nbr	450,800	57,117	135,170	78,041	62,871	201,098

To assist business experts in their interpretation of weak signals, the evolution of these accounts can also be tracked using a Sankey diagram visualization (Fig. 16). In this diagram, the x-axis represents the six episodes, and the y-axis represents the number of times the account appears in the orbits, with the legend at the right captioning the six tracked user accounts.

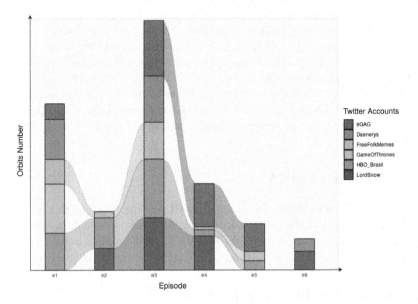

Fig. 16. Sankey diagram showing the evolution of users participating in weak signals, over the six episodes of GOT.

@GameOfThrones posted 46 tweets (retweeted about 139,000 times) on the day of the first episode to reveal some events and encourage the audience to interact. Then this official account disappeared; its role was limited to ensuring that a large audience watches this last season. @LordSnow does not appear in the weak signals of the first episode but from the second episode to disappear after the fourth, we suppose that this could be related to the fact that this character learns the truth about his biological parents only in the last scene of the first episode, which was considered as a decisive event for the next episodes. From the third episode, we notice the presence of comic accounts @9GAG and @FreeFolkMemes that humorously react to the performance of the characters and the setup of the first two episodes, which brought a significant change in the events compared to the previous season. In the last episode, we notice the disappearance of most accounts, we suppose that the season has lost interest.

6.2 Studying the Robustness of BEAM

In this subsection, we will describe an experiment that allowed us to evaluate the robustness of our BEAM method. It consists in reusing the same GOT dataset while varying the interaction graph or the type of nodes. For this purpose, we chose the co-occurrence of hashtags to study their role in weak signals.

The producers of Game Of Thrones have collaborated with Twitter to release hashtags and emojis for twenty main characters. Figure 17 is an extract of these hashtags and their emoji shared by @TwitterTV[13].

We studied the weak signals in the graph of hashtag co-occurrences for the snapshot from 5:00 p.m. to 5:10 p.m. of the first episode. We identified twenty-five precursors, and five weak signal graphlets are given in the Table 7 where the characteristics of the global graph and that of the studied snapshot are also presented. $|V|_G$ and $|E|_G$ are the number of nodes and links in the graph corresponding to the studied period, and $|V|_S$, $|E|_S$ represent the number of nodes, and links in the studied snapshot.

Table 7. Five weak signal graphlets detected in the snapshot from 5:00 p.m. to 5:10 p.m.

$|V|_G$	$|E|_G$	$|V|_S$	$|E|_S$	Weak signal graphlets
503	2353	159	661	G_{19} ⋮ G_{27} ⊠ G_8 △ G_2 △ G_{12} ⌂

We noticed the presence of closed graphlets in the list of detected signals. This presence can be explained by the use of several hashtags in the same tweet.

---

[13] Article published on *Independent.ie*, a news site in Ireland: https://www. independent.ie/style/celebrity/celebrity-news/game-of-thrones-and-twitter-team-up-ahead-of-eighth-and-final-season-37866130.html.

Fig. 17. Hashtags and emoji of Game of Thrones characters.

We used cross-correlation to study the relationship between graphlets' time series and the hashtag co-occurrence time series over the study period. The results of the correlograms can neither confirm nor discard the presence of weak signals in this set. This is due to the critical difference between the number of hashtag co-occurrences and the number of graphlets: some have a very high number, like G_{12} and G_{19}, others like G_{27} are negligible compared to the co-occurrences number. This study shows that using a statistical method is not always sufficient to confirm weak signals, and its results depend mainly on the type of studied data.

In weak signals, we found hashtags corresponding to main characters, in addition to `#GameofThrones`, which was the most frequently used hashtag according to Headline Planet[14] We also found misspelled hashtags like `#GameofThornes` or `#GameofThones`. Table 8 shows three co-occurrents of `#GameofThrones` with the frequency of their appearance together in a single tweet during the 5 pm to 5:10 pm snapshot. In this table, `#filu2019` refers to a hashtag used by a national education organization in Mexico, along with `#facebookdown` and `#domingoderamos`, which signifies Palm Sunday that took place on April 14, 2019. The fourth row of this table regroups hashtags about the main charac-

---

[14] Article in Headline Planet: https://headlineplanet.com/home/2019/04/15/game-of-thrones-has-9-biggest-twitter-trends-following-season-8-premiere/.

ters of the series, and the last one expresses the countdown for the first episode (#conteregresivo), with a play on the words domingo (which means Sunday in Spanish) and got (#domingot).

Table 8. List of the top three co-occurrents of #GameofThrones with their frequency of appearance.

Hashtag1	Hashtag2	Hashtag3	Hashtag4	Frequency
GameofThrones	filu2019	facebookdown	domingoderamos	**81**
	facebookdown	domingoderamos	14deabril	**81**
	domingoderamos	clasicojoven	14deabril	**81**
	daenerystargaryen	cerseilannister	aryastark	3
	forthethrone	domingot	conteregresivo	15

We noticed that in this table, the most frequent co-occurrent hashtags (first three rows) were used to draw attention to other topics on the same day as the episode, which may interest the audience. We also found the hashtag #augmentedreality used to announce Snapshat's first augmented reality lens displaying an ice dragon along a building in New York City (cf. the account @JeffMillerTime carrier of Weak Signal and the second tweet in Fig. 14). This use of hashtags that are not directly related to the series indicates that some users have taken advantage of the popularity of Game of Thrones to give their tweets more exposure.

The hashtag #epicsundays was used by @TheMasters (an account identified as a carrier of weak signal in Sect. 5.3) to announce the Golf Tournament Finals on Sunday, April 14, the day of the first episode airing. Humorously, @TheMasters notes that although the tournament is important, many people will watch GOT. This hashtag was then used by all GOT and golf fans, like the tweet in Fig. 18, where the user uses emojis to symbolize the two events with their time.

Joseph Nezezon
@Joe_P_Nez

🐺 🦌 @ 9:20 AM
🦇 🦊 🦁 @ 9:00 PM

#EpicSundays

12:59 PM · Apr 14, 2019 · Twitter for iPhone

Fig. 18. Tweet using GOT and Golf final tournament hashtags.

To confirm the membership of the hashtags identified in the tweets of users participating in the weak signals of the retweet relationship described in Sect. 5.2, we queried the relational database with the list of hashtags and accounts retrieved in the tweets published around 5 pm. Among the list of accounts, we found @GameofThrones, @TylerIAm, @skyatlantic, etc. Figure 19 illustrates the presence of the hashtags above in the tweets published by users appearing in weak signals. We noticed that most of the users employ the hashtag #GameofThrones and that the hashtag #jonsnow was not used in any of the original tweets of these accounts.

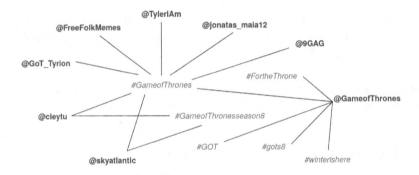

Fig. 19. Relationship between hashtags and users appearing in weak signals.

This study confirmed our hypothesis that the hashtags identified in this section are used in the tweets posted by users appearing in the orbits of weak signals detected in the retweets relation.

Conclusion. The eighth and final season of the Game of Thrones series was a massive success despite mixed feedback and public criticism of its finale. The third episode, seen by a record 17.8 million HBO subscribers, has been called by many observers a defining moment in television history. It was also the episode where we identified the most significant number of weak signals. These results confirmed BEAM's ability to identify, a few hours before the broadcast of each episode, recurrent users (i.e., appearing in several episodes) as carriers of weak signals and, through different visualizations, allow analysts to interpret them.

7   Discovering Latent Variables that Participate the Most in the Signal

We are currently working on a complementary proposal where we aim to detect the presence of latent variables that participate the most in the weak signal. It consists of filtering graphlets to study using a quantitative method based on linear algebra and clustering. The interpretation step remains applicable. In the following, we will present the principles of this proposal and the first results of our experiments and compare them with those obtained in Sect. 5.2.

We choose the singular value decomposition (SVD) [16,47], a linear algebra method, to reduce the data size and highlight singular values participating the most in the information. SVD is used in many applications because it maintains the most significant features of a $m \times n$ matrix by using a smaller matrix size.

To get a better approximation of an original matrix of rank r after applying the SVD, only a few singular values must be retained, while the other singular values are eliminated. All singular values are arranged in descending order on the diagonal of the D matrix. The elimination results from the fact that the first singular value contains the most significant amount of information and that the following singular values contain decreasing amounts of information. These values can therefore be discarded after the SVD is executed, which simultaneously avoids significant distortion of the matrix [2,10]. In this case, the dimensionality of the data representation is reduced to rank p, and essential "latent" variables that exist but are not evident in the representation of the original matrix will be captured. The resulting diagonal matrix is called D_p. The U and V matrices receive the same cutback. The U_p reduced matrix is obtained by removing $r - p$ columns from the U matrix, and the V_p reduced matrix is obtained by eliminating $r - p$ rows from the V matrix.

The choice of the best p rank approximation of a matrix is a compromise between the size of the matrix in terms of columns and rows and the semantics behind the needed approximation. On the one hand, a reduced matrix of $p = 1$ rank (which means that the U and V matrices are also reduced to rank 1) could eliminate essential variables to the analysis of the expert. On the other hand, if p is large enough, the resulting matrix will contain the same volume of information as the original matrix, resulting in the same calculation time as the complete SVD. In both cases, the resulting matrix is not the best approximation of the original matrix. According to Cao et al. [11], the selected value of p must be less than the number of columns in the original matrix. Different methods and algorithms have been proposed to calculate the best value of p. For example, in a paper applied to image compression [2], the authors suggest an upper limit restriction of p to reduce storage space when compressing images with SVD. In particular, they considered the following limit: $p < \dfrac{m \times n}{m + n + 1}$, with m and n, respectively, being the number of rows and columns of the original matrix. According to the different rules considered to calculate the approximation value, we have chosen a scale of values of p, which allows us to better approximate the original matrix without losing the essential variables for further use. We then consider the following limits: $1 < p < \dfrac{m \times n}{m + n + 1}$, with m and n, respectively, being the number of rows and columns.

We then apply the K-means algorithm [32], one of the most known clustering techniques, to identify coherent and distinct subgroups or clusters in the data. We use the elbow method to determine the optimal number of clusters calculated by the algorithm [27]. Figure 20 resumes the following steps in this proposition.

We will describe the experiment on the matrix **User** $\times$ **Graphlet**, corresponding the number of times a user appears in the thirty graphlets for the studied snapshot from 5 pm to 5:10 pm. In this matrix, the rows correspond to the user account identifiers and the columns to the thirty graphlets. Applying the SVD on this matrix, breaks it down into three new matrices: U matrix of rows, i.e. users, of size equal to $8,714 \times 8,714$, V matrix of columns, i.e., graphlets, which size is equal to 30×30, and finally, the diagonal matrix D of singular values of size equal to $8,714 \times 30$.

Therefore, we fix p to 3 in this experiment. According to the fixed value of p, matrix V is reduced to matrix V_p of size 30×3, on which K-means is applied.

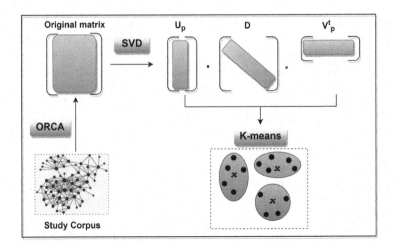

Fig. 20. Summary of the steps followed to detect latent variables from the original matrix.

The same principle applies to matrix U, which is reduced to matrix U_p of size $8{,}714 \times 3$.

The optimal number of clusters returned by elbow is equal to 4. To associate the clustering results with those obtained with BEAM, we apply the K-means on the V_p reduced matrix, corresponding to the graphlets. We iterated several runs, meaning the algorithm returned 4 clusters of graphlets at each run. Table 9 resumes the detected clusters in 4 successive K-means runs. Before each run n, we observe the clusters obtained in the previous run $n-1$. If we find isolated graphlet clusters or clusters of two or three graphlets in this last one, these graphlets are eliminated in the next run n, and K-means is applied to the remaining graphlets. For example, in run 1, we notice that G_4, G_{10} and G_{11} are found isolated in clusters, therefore they are eliminated in run 2, then their features are calculated (velocity, acceleration and contribution). This process aims to reduce the number of graphlets to be computed by BEAM and then verify if isolated graphlets play a significant role in representing weak signals.

We compared the obtained results with Table 4. We noticed that some of the detected weak signal graphlets by BEAM, G_9, G_{12}, G_6 and G_{13} were found isolated in clusters of sizes equal to one or two at most.

We then calculate each run's velocity, acceleration and contribution of isolated graphlets. The results indicate that the weak signal graphlets tend to isolate themselves, or be grouped with another graphlet with a similar shape. Therefore, these isolated graphlets can be considered potential weak signals. We have proposed this method to filter the list of graphlets to be computed by BEAM, instead of calculating all 30 types. However, it cannot replace BEAM since part of the successive iterations of K-means is done manually. Instead, after filtering graphlets, we can examine their feature values, distance to other graphlets of the same cluster or other relevant factors to evaluate their significance.

Table 9. Results of four successive runs of K-means algorithm, by eliminating isolated graphlets.

Iteration	Cluster 1	Cluster 2	Cluster 3	Cluster 4
Run 1	G_4	G_{10}	G_{11}	Remaining graphlets
Run 2	G_{14}	G_1, G_{16}	G_9	Remaining graphlets
Run 3	G_{12}	Remaining graphlets	G_3	G_{17}
Run 4	G_{20}	G_{13}	Remaining graphlets	G_6

We applied SVD with K-means (only with a single iteration) on the user's U_p matrix. The results were consistent with BEAM's; users appearing in weak signals tend to group in the same cluster. In conclusion, the obtained results are promising, and we have illustrated their relevance according to those obtained with the BEAM detection step.

8 Conclusions and Future Works

In this paper, we have outlined our BEAM method that relies on network topology to uncover quantifiable properties that may be characteristic of the weak signal. This is why we have chosen graphlets as an operational tool to detect weak signals. Indeed, graphlets meet the characteristics of weak signals: they are small fragments or patterns of a graph that are not very obvious and of little apparent utility. First, we calculate in a graph of temporal interactions between entities, the **number** of graphlets that represent clear and observable facts, quantifiable with the help of **diffusion** and **amplification** measures that characterize them as precursors. Then, we estimate the signal's **rarity** by measuring the contribution of the graphlets in weak signals. This ratio calculation is used to eliminate false alarms (false positives) and qualify true positives and false negatives as weak signals. Then, we try to confirm the detected weak signal graphlets with the correlation between the graphlet time series and the data of the study corpus. Once the weak signals have been detected and confirmed, experts must decide on their relevance for the organization. Here we move from a quantitative stage (detection) to a qualitative stage (interpretation), where the role of the experts is crucial. Lastly, we presented a case study on the Game of Thrones data, one of the datasets we used in our experiments. The experiments that we carried on had various goals. We first demonstrated BEAM's ability to detect and interpret weak signals concerning ground truth. Our second objective was to study the reproduction of results with periodic data. Here, the dataset covers the retweets produced on the broadcast of the six episodes of the last season of the TV series Game of Thrones (GOT), presented in the case study in this paper. In addition, we wanted to examine the method's sensitivity while varying the experimental conditions. In this paper, we also presented the GOT dataset by changing the type of nodes and studying the space of hashtags. Finally, we validated our method with opposing examples of scheduled/repetitive events and other cases where the study context was unknown.

The results of our experiments supported our hypothesis that graphlets can be considered a signature of a weak signal. They allow us to automate the task of detecting weak signals in a large volume of data while leaving room for expert interpretation, thus eliminating the black box effect that a fully automated method could have.

In further research, we would like to enlarge our scope and explore the use of clustering techniques for the discovery of groups and profiles of nodes in the studied temporal graphs. We also seek to enrich the results of our method by adding ontological insights and sentiment analysis to study threats and opportunities in more detail. From an applied point of view, we want to offer business experts tools to help them in decision-making (a practical example can be argumentation diagrams). Lastly, we seek to extend the method to detect phase transitions corresponding to sudden state changes in an environment.

Acknowledgements. This work is supported by the Investissements d'Avenir program, Cocktail ISITE-BFC project (Initiatives Science Innovation Territoires Économie en Bourgogne-Franche-Comté), ANR contract 15-IDEX-0003, https://projet-cocktail. fr/.

References

1. Aiello, L.M., et al.: Sensing trending topics in Twitter. IEEE Trans. Multimedia **15**(6), 1268–1282 (2013)
2. Aishwarya, K.M., Ramesh, R., Sobarad, P.M., Singh, V.: Lossy image compression using svd coding algorithm. In: 2016 International Conference on Wireless Communications, Signal Processing and Networking (WiSPNET), pp. 1384–1389. IEEE (2016)
3. Ansoff, H.I.: Managing strategic surprise by response to weak signals. Calif. Manag. Rev. **18**(2), 21–33 (1975)
4. Ansoff, H.I., McDonnell, E.J.: Implanting strategic management (1990)
5. Baiesi, M.: Scaling and precursor motifs in earthquake networks. Physica A **360**(2), 534–542 (2006)
6. Blanco, S., Lesca, N.: From weak signals to anticipative information: learning from the implementation of an information selection method. In: Proceedings of the International Conference In Search of Time (ISIDA), pp. 197–210. Citeseer (2003)
7. Blondel, V., Guillaume, J.-L., Lambiotte, R., Lefebvre, E.: Fast unfolding of communities in large networks. J. Stat. Mech: Theory Exp. **2008**(10), P10008 (2008)
8. Brewster, B., Andrews, S., Polovina, S., Hirsch, L., Akhgar, B.: Environmental scanning and knowledge representation for the detection of organised crime threats. In: 21th International Conference on Conceptual Structures (ICCS), July 2014
9. Brin, S., Page, L.: The anatomy of a large-scale hypertextual web search engine. Comput. Netw. ISDN Syst. **30**(1–7), 107–117 (1998)
10. Burden, R.L., Faires, J.D.: Numerical Analysis, 9th edn. Brooks Cole Cengage Learning, Boston (2011)
11. Cao, L.: Singular value decomposition applied to digital image processing. In: Division of Computing Studies, pp. 1–15. Arizona State University Polytechnic Campus, Mesa, Arizona State University polytechnic Campus (2006)

12. Chatfield, C.: The Analysis of Time Series: An Introduction. Chapman and Hall/CRC, Boca Raton (2003)
13. Coffman, B.: Weak signal research, part I: introduction. J. Transit. Manag. **2**(1) (1997)
14. Davies, T., Marchione, E.: Event networks and the identification of crime pattern motifs. PLoS ONE **10**(11), e0143638 (2015)
15. Dotsika, F., Watkins, A.: Identifying potentially disruptive trends by means of keyword network analysis. Technol. Forecast. Soc. Chang. **119**, 114–127 (2017)
16. Eckart, C., Young, G.: The approximation of one matrix by another of lower rank. Psychometrika **1**(3), 211–218 (1936)
17. El Akrouchi, M., Benbrahim, H., Kassou, I.: End-to-end LDA-based automatic weak signal detection in web news. Knowl.-Based Syst. **212**, 106650 (2021)
18. Friendly, M.: Corrgrams: exploratory displays for correlation matrices. Am. Stat. **56**(4), 316–324 (2002)
19. Godet, M.: From Anticipation to Action: A Handbook of Strategic Prospective. UNESCO Publishing, Paris (1994)
20. Goldin, D.Q., Kanellakis, P.C.: On similarity queries for time-series data: constraint specification and implementation. In: Montanari, U., Rossi, F. (eds.) CP 1995. LNCS, vol. 976, pp. 137–153. Springer, Heidelberg (1995). https://doi.org/10.1007/3-540-60299-2_9
21. Hiltunen, E.: The future sign and its three dimensions. Futures **40**(3), 247–260 (2008)
22. Hiltunen, E.: Weak signals in organizational futures learning. Ph.D. thesis, Helsinki School of Economics (2010)
23. Hočevar, T., Demšar, J.: A combinatorial approach to graphlet counting. Bioinformatics **30**(4), 559–565 (2014)
24. Ilmola, L., Kuusi, O.: Filters of weak signals hinder foresight: monitoring weak signals efficiently in corporate decision-making. Futures **38**(8), 908–924 (2006). Organisational Foresight
25. Jordan, J.: The Four Stages of Highly Effective Crisis Management: How to Manage the Media in the Digital Age. CRC Press, Boca Raton (2011)
26. Juszczyszyn, K., Kołaczek, G.: Motif-based attack detection in network communication graphs. In: De Decker, B., Lapon, J., Naessens, V., Uhl, A. (eds.) CMS 2011. LNCS, vol. 7025, pp. 206–213. Springer, Heidelberg (2011). https://doi.org/10.1007/978-3-642-24712-5_19
27. Kodinariya, T.M., Makwana, P.R.: Review on determining number of cluster in k-means clustering. Int. J. **1**(6), 90–95 (2013)
28. Korvers, P.M.W.: Accident Precursors: Pro-active Identification of Safety Risks in the Chemical Process Industry. eLibrary.ru (2004)
29. Kuosa, T.: Futures signals sense-making framework (FSSF): a start-up tool to analyse and categorise weak signals, wild cards, drivers, trends and other types of information. Futures **42**(1), 42–48 (2010)
30. Kwon, L.-N., Park, J.-H., Moon, Y.-H., Lee, B., Shin, Y.H., Kim, Y.-K.: Weak signal detecting of industry convergence using information of products and services of global listed companies - focusing on growth engine industry in South Korea. J. Open Innov. Technol. Mark. Complex. **4**(1) (2018)
31. Lesca, H., Lesca, N.: Weak Signals for Strategic Intelligence: Anticipation Tool for Managers. ISTE Ltd. (2011)
32. Lloyd, S.: Least squares quantization in PCM. IEEE Trans. Inf. Theory **28**(2), 129–137 (1982)

33. Macrae, C.: Interrogating the unknown: risk analysis and sensemaking in airline safety oversight. Number 43 in CARR Discussion Papers (DP 43). Centre for Analysis of Risk and Regulation, London School of Economics (2007)

34. Maitre, J., Ménard, M., Chiron, G., Bouju, A., Sidère, N.: A meaningful information extraction system for interactive analysis of documents. In: International Conference on Document Analysis and Recognition (ICDAR), pp. 92–99 (2019)

35. McMinn, A.J., Moshfeghi, Y., Jose, J.M.: Building a Large-scale corpus for evaluating event detection on Twitter. In: Proceedings of the 22nd ACM International Conference on Information and Knowledge Management, CIKM '13, pp. 409–418. Association for Computing Machinery, New York, NY, USA (2013)

36. Mendonça, S., Cardoso, G., Caraça, J.: The strategic strength of weak signal analysis. Futures **44**(3), 218–228 (2012)

37. Mühlroth, C., Grottke, M.: A systematic literature review of mining weak signals and trends for corporate foresight. J. Bus. Econ. **88**(5), 643–687 (2018). https://doi.org/10.1007/s11573-018-0898-4

38. Phimister, J., Bier, V., Kunreuther, H.: Accident Precursor Analysis and Management: Reducing Technological Risk Through Diligence. The National Academies Press, Washington, DC (2004)

39. Pržulj, N., Corneil, D.G., Jurisica, I.: Modeling interactome: scale-free or geometric? Bioinformatics **20**(18), 3508–3515 (2004)

40. Rascão, J.P.: Strategic information surveillance. In: Leon, R.-D. (ed.) Managerial Strategies for Business Sustainability During Turbulent Times, pp. 78–99. IGI Global (2018)

41. Ribeiro, P., Paredes, P., Silva, M.E.P., Aparicio, D., Silva, F.: A survey on subgraph counting: concepts, algorithms and applications to network motifs and graphlets. arXiv preprint arXiv:1910.13011 (2019)

42. Rossel, P.: Early detection, warnings, weak signals and seeds of change: a turbulent domain of futures studies. Futures **44**, 229–239 (2012)

43. Rousseau, P., Camara, D., Kotzinos, D.: Weak Signal Detection and Identification in Large Data Sets: A Review of Methods and Applications. Springer, Cham (2021). https://doi.org/10.13140/RG.2.2.20808.24327/1

44. Saritas, O., Smith, J.: The big picture - trends, drivers, wild cards, discontinuities and weak signals. Futures **43**, 292–312 (2011)

45. Schoemaker, P.J.H., Day, G.S.: How to make sense of weak signals. Leading Organ. Perspect. New Era **2**, 37–47 (2009)

46. Soufiani, H.A., Airoldi, E.: Graphlet decomposition of a weighted network. In: Lawrence, N.D., Girolami, M. (eds.) Proceedings of the Fifteenth International Conference on Artificial Intelligence and Statistics, Proceedings of Machine Learning Research, vol. 22, pp. 54–63, La Palma, Canary Islands, 21–23 April 2012. PMLR (2012)

47. Stewart, G.W.: On the early history of the singular value decomposition. SIAM Rev. **35**(4), 551–566 (1993)

48. Tang, J.: Computational models for social network analysis: a brief survey. In: Proceedings of the 26th International Conference on World Wide Web Companion, WWW '17 Companion, Republic and Canton of Geneva, CHE, pp. 921–925. International World Wide Web Conferences Steering Committee (2017)

49. Thorleuchter, D., Van den Poel, D.: Idea mining for web-based weak signal detection. Futures **66**, 25–34 (2015)

50. van Veen, B.L., Ortt, J.R.: Unifying weak signals definitions to improve construct understanding. Futures **134**, 102837 (2021)

51. Venugopal, V., Ates, A., McKiernan, P.: Revisiting Ansoff's weak signals - a systematic literature review. In: 36th Annual Conference of the British Academy of Management (2022)
52. Welz, K., Brecht, L., Pengl, A., Kauffeldt, J.V., Schallmo, D.R.A.: Weak signals detection: criteria for social media monitoring tools. In: ISPIM Innovation Symposium, p. 1. The International Society for Professional Innovation Management (ISPIM) (2012)
53. Yoon, J.: Detecting weak signals for long-term business opportunities using text mining of Web news. Expert Syst. Appl. **39**(16), 1243–1250 (2012)

Engineering Runtime Root Cause Analysis of Detected Anomalies

Zisis Flokas$^{(\boxtimes)}$ and Anastasios Gounaris

Aristotle University of Thessaloniki, Thessaloniki, Greece
{flokaszisis,gounaria}@csd.auth.gr

Abstract. The main objective of this work is to provide a unified, easy to configure and extensible end-to-end system that performs root cause analysis (RCA) methods on top of anomaly detection (AD) methods in an online setting. AD-focused RCA for online settings has not been investigated so far; therefore our work can be seen as an initial approach to this end. Inspired by the solutions developed in the ThirdEye project, which is coupled with the Apache Pinot data warehousing system, we re-engineer ThirdEye's RCA components/techniques so that they operate in a manner that they can directly ingest input records from Apache Kafka and continuously compute aggregates at different level of granularity in a principled manner for both OLAP queries and provision of baselines to support RCA. To attain scalability, we build our solution in the Apache Flink stream processing engine. This work presents the main design choices when applying ThirdEye's concepts on data streams and presents indicative examples and scalability experiments. Our solution is provided in open-source.

Keywords: root cause analysis · anomaly detection · data streams · Flink · Kafka

1 Introduction

Data warehousing is a mature technology relying on the data cube logical model and OLAP operations [24]. Several modern tools, such as Apache Pinot [5], provide scalable OLAP solutions, whereas automated extraction of useful insights, such as anomalies, is a hot topic [31]. However, many modern solutions require online processing of the data records before these are loaded to data cubes. Anomaly detection (AD) over data streams is a mature field [10,39] and there exist multiple solutions to this end. However, the issue of combining continuous AD with online root cause analysis (RCA) is overlooked [35]. Thus, the main motivation of this work is to fill this gap and explain how to develop a runtime RCA method for continuous AD through re-engineering a solution that performs AD-oriented RCA on top of data cubes.

© The Author(s), under exclusive license to Springer-Verlag GmbH, DE, part of Springer Nature 2023
A. Hameurlain and A. M. Tjoa (Eds.): *Transactions on Large-Scale Data-and Knowledge-Centered Systems LV*, LNCS 14280, pp. 64–86, 2023.
https://doi.org/10.1007/978-3-662-68100-8_3

More specifically, in this work, we are interested in a particular solution originally developed by LinkedIn, namely ThirdEye [42]. LinkedIn has open-sourced several important platforms, including the well-established Apache Kafka [4] event processing and Apache Pinot [5] data warehousing ones. The distinctive feature of ThirdEye is that it couples AD with RCA. However, this RCA module operates only in a batch mode after data have been loaded to a Pinot data warehouse.

The aim of this work is to re-engineer ThirdEye's RCA components and techniques so that they operate in a manner that they can directly ingest input records from Apache Kafka. This has several implications, such as that several statistical metadata need also to be computed on the fly thus raising scalability issues. Data cubes come with hierarchical dimensions, so that aggregate measurements can be extracted at numerous levels of granularity through OLAP operations. Baseline statistics are also required to perform RCA. Therefore, in an online setting that combines AD and RCA, multiple current and baseline aggregates need to be computed in an efficient manner.

The contribution of this work is the presentation of the detailed engineering approach followed so that input records are aggregated at multiple levels along with their baseline values; on top of such aggregates, with lightweight runtime post-processing, both AD and RCA can be performed on the fly. To attain the required scalability, we build our solution in the Apache Flink stream processing engine. The proposed system along with instructions to use it, is publicly available as a GitHub project [18]. The solution is extensible by design; e.g., any runtime AD module can be encapsulated.

The remainder of this work is structured as follows. In the next section, we discuss the related work. The background on ThirdEye is provided in Sect. 3. Our solution is presented in Sects. 4 and 5, where the infrastructure and all the steps in the pipeline are explained in detail, respectively. In Sect. 6, we present scalability experiments, and we conclude in Sect. 7.

2   Related Work

There have been numerous studies of AD recently, e.g., [10,11,21,28,34,36,39, 45]. In time series, techniques are distinguished according to their input data, the nature of the method and the outlier type. Input data can be univariate or multivariate, while methods can be unsupervised, semi-supervised or supervised. Finally, outlier types in time series comprise point anomalies (either contextual or not), collective anomalies that refer to anomalous subsequences, or anomalous time series, where a complete time series may be characterized as an anomaly [10,36,39].

Unsupervised methods do not require a training step. They identify anomalies by studying the shape, frequency or distribution of input data points, with a view to detecting the points that deviate significantly from normal ones. This significance can be something that a model adapts to or is provided by the user. Semi-supervised techniques try to model the normal behavior as they are trained solely on normal data points; when they come across new data they can classify them according to the normal model or highlight an anomaly. Supervised techniques require a training step in order to model normal and abnormal behavior for data points in a time series. At a lower level, in [39], anomaly detection methods for time series are classified into six families, namely forecasting-, reconstruction-, encoding-, distance-, distribution-, and isolation tree-based ones.

The techniques we employ in our solution are univariate ones, since each data cube measure is treated individually to reduce complexity. In this filed in the literature, there are some basic estimation models that draw their ideas from basic statistics, such as [9,32] to produce expected values. Other techniques assume a distribution or a fitted model that generates data and point out the unlikely ones. Examples of such methods using smoothing techniques are [12,15, 25], Exponentially Weighted Moving Average [13,40,53] and Gaussian Mixture Model-based solutions [37]. Another popular family of methods in literature is the density-based models. These methods measure abnormality by counting the number of neighbors of each record; if the neighbors are less than a threshold value, then the record is annotated as an anomaly. Neighbors are considered the data points in distance no larger than a radius value R. This is a category of methods that is well studied also in an online setting using sliding windows, e.g., [2,27,34,48].

RCA combined with AD, being closely related to explainability, is a topic that has received an increasing attention in the past years. One of the early attempts to systematically review RCA methods that are AD method-agnostic is presented in [33], where both point explanation [30,51] and summary explanation techniques [23,29] are discussed. Our setting relates to point explanation techniques. The RefOut [30] algorithm randomly samples a given number of subspaces with a fixed dimensionality to find the most relevant ones. Beam [51] also studies subspaces up to a given dimensionality, where all possible $2d$ subspaces are scored. Two of the most advanced techniques are EXstream [54] and MacroBase [1]. The former refers to a CEP (Complex Event Processing) setting but for the RCA part it resorts to archived streams; i.e., it is not suitable for fully runtime solutions. Macrobase can be used to perform AD and RCA but includes computationally demanding operations such as FPGrowth. A survey on outlier explanations is provided at [35], where it is shown that techniques tailored to data streams have not received adequate attention, and as explained previously, event those that target data streams, they cannot apply RCA at runtime.

In summary, none of the above solutions is designed to operate in a real-time streaming setting both for the AD and the RCA part. Similarly, established techniques from the XAI field, such as [22,38], are not suitable for a runtime environment. An exception is the work in [49], which considers distance-based point anomalies and is capable of tagging each identified outlier as being an isolated point, or a point near a cluster, or a point belonging to a micro-cluster and so on. In other words, it is capable of deriving runtime explanations but fails to perform RCA, for which identification of the dimensions causing the anomalous behavior is required. As such, our work can be deemed as an initial attempt to render RCA for AD applicable on the fly.

3   Background: ThirdEye's AD and RCA Approach

In this section, we provide an overview of the AD and RCA functionality of ThirdEye, as reported in the StarTree ThirdEye documentation [42] both for the enterprise and community editions [44], as well as in the open-sourced project of the community edition [43] and the documentation of the initial (and now archived) ThirdEye project [47].

ThirdEye comprises a *batch* AD rule-based system, the detection process of which is triggered using cron-based schedules. The default cron schedule is determined based on the data granularity of the examined data cube stored in Pinot [5]. For each cron schedule triggered, data from the time period between current and previous triggered detection processes are fetched from the Pinot data warehouse to perform the analysis. Rule-based solutions are considered to be simple ones, but according to [42], AD needs to be simple in logic to facilitate reasoning. ThirdEye (in the enterprise edition) provides, apart from several built-in AD rules, the option to expose a time series metric to an external HTTP service to perform the AD. In such a manner, more advanced AD solutions can be employed.

ThirdEye currently works only with univariate time series input. The AD methods provided rely on either thresholds or past data to define the "expected" (a.k.a "baseline") value. This allows us to transfer all these detection rules in an online streaming setting as decisions can be taken on the fly by creating processes to define and maintain state/window of past data. However, this process needs to be conducted in an efficient manner since, in data warehouses, the aggregated information can be provided at multiple levels of granularity. The detection rules include ones that compute the absolute or relative difference of the current value from the baseline (using also traditional statistics such as mean and standard deviation) and algorithms, such as Holt-Winters [26,52].

Additionally to the detection rules, ThirdEye provides anomaly filters in order to filter out anomalies detected based on different rules. In this way, more complex detection pipelines can be created as multiple detection rules and anomaly filters can be combined in a single detector.[1] Examples include filtering the anomalies based on the anomaly duration, or, if compared to the baseline, absolute change is below a certain threshold. Finally, ThirdEye provides the functionality to merge anomalies detected in sequence to provide more concise results to the user.

Regarding RCA, ThirdEye provides two main techniques. The first method is called "Heatmap" while the latter is called "Top Contributors (TC)" formerly named "Data Cube Algorithm (DCA)" [14]. The Heatmap method provides a means to visualize how a metric sliced by each dimension value (at any level of the hierarchy) changes when compared to a baseline. By default, the baseline is set to one week before the anomaly event took place. There are multiple ways to measure changes based on simple mathematical formulas (e.g., percentage change, change in contribution, relative contribution to overall change).

---

[1] This is a feature deprecated in StarTree version of ThirdEye (but still available in the archived version).

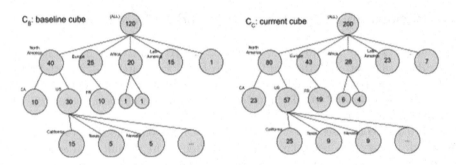

Fig. 1. The metric breakdown of baseline and current data cubes with continent, country, and state dimension values (from [14,41])

The TC algorithm[2] has a different focus and points out which are the dimensions that contribute the most to an anomaly detected. While Heatmap provides a nice way to visually inspect how the dimension values of each metric contribute to the anomaly detected, it is hard to follow this manual exploration process especially when the metric observed is accompanied by multiple dimensions. A data cube may contain 10 to 20 dimensions, which correspond to millions of data segment combinations. The algorithm constructs two graphs based on data cube hierarchies. An example is provided in Fig. 1, which shows the dimension breakdown of these cubes. Note that this figure has omitted some nodes due to space constraints. For instance, we only show FR as a child of Europe with the children of FR omitted.

The impact of the change is calculated in a node (of the graph) with three factors: change ratio, change difference, and segment contribution. Intuitively, change ratio measures how big the change is. Change difference measures the unexpected change compared to its parents. In other words, it computes how much surprise the change is compared to its parent's change. The segment contribution measures the change significance score of each node, given the baseline and current values of a node and its parent. More formally, given the baseline and current values of a node n and its parent, the change significance score is calculated as:

$$significance(n) = (vC - r \cdot vB) \cdot log\left(\left(\frac{vC}{r * vB} - 1\right) \cdot \left(\frac{contributionC}{contributionAll}\right) + 1\right) \quad (1)$$

where vB and vC are the baseline and current node values, respectively; r is the expected change ratio between the baseline and current from its parent node,

---

[2] It is important to note that in [41], it is stated that the original DCA is part of the enterprise edition of the StarTree ThirdEye, while for the community edition, there is a simpler algorithm that does not consider multiple dimensions jointly when making the analysis.

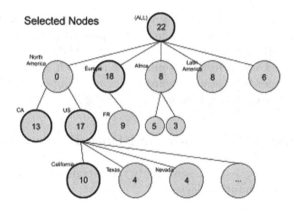

Fig. 2. The delta change and selected nodes between the data cubes (from [14,41])

| Heatmap | Table | Algorithm |

☼ Configure analysis table

Top Anomalous Dimensions | | | | Overall Change -18.56%

dimension1	dimension2	dimension3	Current/Baseline	Contribution to Overall Change		% Change	Change in Contribution
▣ Other	All	All	24,707 / 28,479	████	-36.9189%	-13.2448%	3.3766%
▣ CN	Other	value1	2,334 / 1,987	▌	3.3963%	17.4635%	1.5966%
▣	cn	value2	1,710 / 4,192	███	-24.2928%	-59.2080%	-3.8008%
▣ US	us	value1	1,147 / 2,332	██	-11.5983%	-50.8148%	-1.6778%
▣		value2	1,235 / 2,016	▌	-7.6441%	-38.7401%	-0.9075%
▣		value3	492 / 899	▌	-3.9836%	-45.2725%	-0.5357%
▣	unknown	All	2,258 / 2,698	▌	-4.3065%	-16.3084%	0.1355%
▣ country1	v1	Other	4,074 / 4,404	▌	-3.2299%	-7.4932%	1.0872%
▣		value1	5,819 / 6,621	██	-7.8497%	-12.1130%	0.9522%
▣ country2	v2	All	1,055 / 1,420	▌	-3.5725%	-25.7042%	-0.2263%

Show 1 - 10 of 10 | 25 ⬍ | < ‹ › >

Fig. 3. RCA results representation example (from [14,41])

which is defined as, $r = \frac{v\_parentC}{v\_parentB}$; $contributionC$ is the contribution of the current node; and $contributionAll$ is the overall contribution. For additive metrics, the contribution can simply be calculated as $(vB + vC)$. For ratio metrics, it could be simply defined as $(DenominatorB + DenominatorC + NumeratorB + NumeratorC)$ or calculated from an additional additive metric.

The importance of a dimension value d (e.g., country) is defined as the sum of the significance score of all its children (e.g., (country = "US"), (country = "FR"), and so on). Formally, the dimension importance of d is:

$$Dimension\_importance(d) = \sum_{n=1}^{m} significance(n) \qquad (2)$$

where m is the children count of dimension value d.

Afterwards, the dimension importance is used to determine the tree structure as shown in Fig. 2. The root level is the most important dimension. Finally, each parent node picks the top k (i.e., the summary size) children nodes and we merge the result of parent nodes from bottom to the top; each merge operation keeps only the top k nodes. The results regarding the most important dimensions can be easily summarized in a table or as shown in Fig. 3.

4   Infrastructure Overview and Data Representation

The aim of this section is to present firstly the overall system architecture, and then, the data representation devised.

4.1   Infrastructure Overview

The proposed system is flexible regarding the different types of data sources it can consume and analyze. Data is ingested to the AD and RCA modules through connection with topics of a Kafka Broker, providing a concrete and flexible way to integrate this solution in an existing environment. Kafka is a widely used tool for data integration across heterogeneous systems such as different types of database engines, file systems and search engines, with many out of the box solutions, provided primarily through the Kafka Connect [8,16] subproject. Consequently, the solution of this work can be considered more flexible compared to the ThirdEye project, which strictly requires Apache Pinot tables as input for the StarTree version [42] and additionally MySQL and Presto table sources in its deprecated version [46]. For the data processing part, Apache Flink [3] was chosen as it is also a well established system and framework for native processing of data streams used by Big Tech companies, such as Uber [20].

Overall, the proposed solution consists of two main components, a Kafka broker for streaming data and a Flink cluster for the processing of those streams. Apache Kafka requires using Apache Zookeeper [6] to coordinate its various processes, such as broker discovery, consumer offset management, topic configuration and more. The Flink cluster consists of a Job Manager, which is responsible for the management of resources and the coordination of the life cycle of the jobs running in the cluster, and one or more Task Managers, which are the workers that actually execute the jobs. In this work's setting, Task Managers can be scaled and configured through the Docker Compose configuration that is provided in the GitHub repository [18]. The whole infrastructure is provisioned in the form of containers using the Docker engine [17].

Regarding the platform choices presented above, a main requirement that has to be met is support for scalable real-time processing; here, real-time processing corresponds to processing as soon as data arrive without imposing a hard time constraint. Both Kafka and Flink excel at both scalability and real-time processing support. The scalability is required because continuous RCA inherently relies on the provision of numerous baseline aggregated metrics, i.e., for each incoming record, which is a JSON message in our implementation that includes

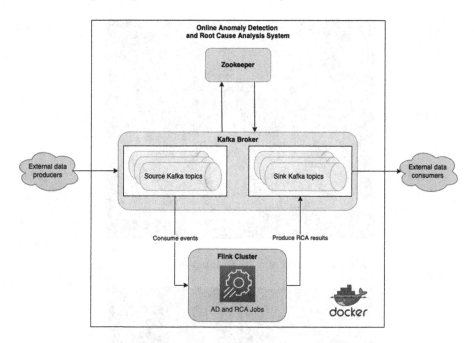

Fig. 4. Infrastructure overview

a metric field, the timestamp and all the dimension hierarchy information, multiple aggregated values need to be updated/computed. The details are provided below. An overview of the infrastructure is provided in Fig. 4.

4.2 Data Representation

Input data records are encoded as JSON messages. The example dataset used in the evaluation (see Sect. 6) comes from the TPC-DS benchmark dataset [50], where each record comprises, apart from its id, a timestamp, 2 measures and multiple dimensions. The dimensions have hierarchies and correspond to the item category, the delivery place and the shipping mode. An example is presented in Fig. 5.

The records arriving from Kafka to Flink are transformed to a format called *InputRecords*, where the dimension hierarchy information is embedded. To this end, we define the *Dimension* tuples, which consist of four attributes, namely, name, value, group and level. The latter attributes are computed with the help of dimension configuration input metadata. The configuration metadata form is shown in Fig. 6. An example of a *Dimension* record is *Dimension*(name = "ca_city", value = "Spring Hill", group = "spatial", level = 3). The schema of *InputRecords* is presented in Table 1.

The input stream builder receives records as shown in Fig. 5 and with the help of the configuration files, as shown in Fig. 6, it transforms them to a stream of *InputRecords*.

```
 1 {
 2   "ws_item_sk" : 13550,
 3   "ws_quantity" : 15,
 4   "ws_ext_list_price" : 1360.2,
 5   "i_brand_id" : 5004001,
 6   "i_class_id" : 4,
 7   "i_category_id" : 5,
 8   "i_manufact_id" : 950,
 9   "ca_city" : "Spring Hill",
10   "ca_county" : "Shelby County",
11   "ca_state" : "IL",
12   "sm_type" : "REGULAR",
13   "sm_code" : "SURFACE",
14   "sm_carrier" : "TBS",
15   "sale_at" : "1998-01-22T18:50:23"
16 }
```

Fig. 5. Example of a JSON input event in Kafka

```
input_stream {
    input_topic = "test1"
    timestamp_field = "sale_at"
    value_field = "ws_quantity"
    dimensions {
        names = ["ca_city", "ca_county", "ca_state", "sm_code"]
        definitions {
            ca_city {
                value_type = string
                parent_dimension = ca_county
                group = spatial
            }
            ca_county {
                value_type = string
                parent_dimension = ca_state
                group = spatial
            }
            ca_state {
                value_type = string
                parent_dimension = root
                group = spatial
            }
            sm_code {
                value_type = string
                parent_dimension = root
                group = delivery
            }
        }
    }
}
```

Fig. 6. Input Stream configuration example

Table 1. InputRecord schema

Field	Scala type	Description
id	String	Unique id for event
timestamp	String	Event timestamp
value	Double	The metric field
dimensions	Map[String, Dimension]	Map of dimensions with dimension name as key
dimensionsHierarchy	Map[Dimension, Dimension]	Map where keys are all the Dimensions that have a parent, each key refers to the parent Dimension
timestampPattern	String	The formatter to use for transforming timestamp to epoch time. Defaults to yyyy-MM-dd HH:mm:ss

5   Implementation Details

We have already described the data representation above. The remainder implementation choices in a step-by-step manner in order to attain an RCA-enabled online AD solution.

5.1   Metric Aggregator

ThirdEye generates the time series of the metric that is being studied for anomalies by performing an aggregation query on the metric column, according to a given aggregation function and a specific time granularity that the user provides. The Metric Aggregator, which is discussed in this section, aims to replicate this process in a streaming setting by fetching input records from Kafka, grouping them using sliding time windows and applying the aggregation functions over these windows.

Metric Aggregator is the process that is applied over a stream of JSON input records transformed to the format of *InputRecords* as discussed above and produces a new stream of aggregated records. The Metric Aggregator initially splits the original stream into chunks using windows. The type of windows used is sliding time-based ones, where, by default, the window size is set to 300 s and the slide is 60 s. While ThirdEye creates non overlapping chunks of data, in the online solution, it was decided to use sliding windows, which fit better in the streaming setting, instead of tumpling ones, as our time chunks are much smaller than those in scheduled batches. Each window slide groups an arbitrary number of input records and the result is represented by a single *AggregatedRecords* record in the new transformed stream. The schema of the *AggregatedRecords* model is provided in Table 2.

Table 2. AggregatedRecords schema

Field	Scala type	Description
current	Double	Aggregated metric value of *InputRecords*
windowStartingEpoch	Long	UTC epoch time of first element in the window, derived from *InputRecord* timestamp field
RecordAggregated	Int	Number of records aggregated in the window
DimensionsBreakdown	Map[Dimension, Double]	Contribution of each dimension in current value
DimensionsHierarchy	Map[Dimension, Dimension]	Hierarchy map of all unique dimensions in the windows

Metric Aggregator can provide different types of aggregators and their respective accumulators based on the kind of aggregation function the user wants to apply over input data, e.g., sum, average, count, count distinct, max, min, 50/90/95/99th percentile and so on. In the remainder of this work, we concentrate only on the sum aggregator. An aggregator is also responsible for generating a map of dimension hierarchies and a map of dimensions breakdown. More specifically, the aggregated record contains all the unique pairs *(ChildDimension, ParentDimension)* found in a window. Apart from the metric and the dimension hierarchies, an *AggregatedRecords* record must also include the participation of each dimension value in the aggregated metric, i.e., the dimension breakdown. In order to generate the dimensions breakdown, for each *Dimension* attribute of an input record, a *(Dimension, value)* tuple is generated. This process creates a map of type *Map[Dimension, Double]* for all dimensions present in the window, where each item represents a node as shown in the current dimension breakdown tree of Fig. 1, except the root node which is actually represented by the current field. In this way, an *AggregatedRecords* record serves as a summary of all the input records found in a window.

The aggregation of the metric is primarily required by the AD module in order to perform the outlier detection while the dimension hierarchies and breakdown attributes are required by the RCA module. If someone needed only to perform anomaly detection without RCA, the aggregator could be simplified.

It is important to note that in ThirdEye, the metric aggregation step is completely decoupled from the dimensions breakdown step, which is required for the RCA part. This happens due to the fact that, in ThirdEye, the source of data for the analysis is a data cube that can be easily accessed to fetch again data when the RCA process is triggered rather than having a single step doing the aggregation job for both the AD and RCA processes. Additionally, this implies that in a batch processing environment, the additional data retrieval and dimensions breakdown generation is done only for anomalies detected and not for every single aggregated time chunk. Decoupling these two processes in a streaming environment would mean that the stream of data should be ingested twice by two separate processes that would aggregate the records using the same windowing approach to perform the different aggregations. This would require adding a new layer that would join those two parallel streams on the fly, which may be possible but is complicated. Finally, we highlight that, in our solution, the RCA rationale runs continuously and not only when anomalies are detected.

5.2   Baseline Aggregator

Baseline Aggregator is responsible for generating the baseline values that are leveraged by (some) anomaly detectors and the RCA module (dimension values baseline). The latter needs to compute dimension value change statistics across different time periods. Even when AD does not rely on baselines, e.g., it is a simple threshold-based technique, RCA is always baseline-driven regardless the AD method used for the detection step. To simplify the interface and satisfy the streaming requirement, in this implementation, Baseline Aggregator performs both metric and dimension values aggregations in a single pass.

The Baseline Aggregator requires as input a stream of *AggregratedRecords* to produce a stream of *AggregatedRecordsWBaseline* records. In the schema of Table 2, the current and dimension breakdown values are complemented with corresponding baseline values. More specifically, as already mentioned, each slide is by default 60 secs (although it is configurable). Comparing a time chunk of 60 s with the time chunk of the previous 60 s seems not to be a good idea as this method is prone to missing out anomalies because an anomaly could last for several minutes and in this case, both baseline and current values would be anomalous. Additionally, this approach could lead to giving fault explanations for the detected anomalies as, in some time chunks, only very few data points are present and there may be cases that the dimension values of two different time periods are very different mostly because not many input records participate in them.

To address these issues, in this work we introduce a count-based (meta-) window of size n, which corresponds to multiple *AggregratedRecords*. The *AggregratedRecords* record with the latest *windowStartingEpoch* timestamp in each window is treated as the current one and the rest of the $n - 1$ records are processed to derive the baseline. The default size of n is set to 10 meaning that current is represented by a single *AggregratedRecords* record and baseline corresponds to 9 accumulated *AggregratedRecords*. As the baseline is now a list of records, an aggregation must be performed to gain a single summary record. We have chosen to use a simple average aggregation to create the baseline record; however, more sophisticated solutions can be readily implemented. For the *baselineDimensionBreakdown*, the dimensions of all records are grouped and then averaged per group, while the *dimensionHierarchy* maps of both current and baseline records are merged.

An example of a *AggregatedRecordsWBaseline* record instance with some dimension values of the dimension breakdown truncated is given in Fig. 7. The complete record is much larger. This example uses the TPC-DS benchmark as previously. Based on this example, we observe that the current value, which is equal to 607, is shared between dimension values BIKE, SURFACE and SEA of the sm_code dimension in the delivery group. If we sum the contribution of these three dimension values, we get the overall current value 607.

```
 1  AggregatedRecordsWBaseline(
 2      current=607.0,
 3      baseline=656.0,
 4      current_dimensions_breakdown=Map(
 5          Dimension(name=ca_state, value=MO, group=spatial, level=1) -> 607.0,
 6          Dimension(name=ca_city, value=Woodville, group=spatial, level=3) -> 607.0,
 7          Dimension(name=sm_code, value=SEA, group=delivery, level=1) -> 381.0,
 8          Dimension(name=sm_code, value=BIKE, group=delivery, level=1) -> 51.0,
 9          Dimension(name=ca_county, value=Shannon County, group=spatial, level=2) -> 607.0,
10          Dimension(name=sm_code, value=SURFACE, group=delivery, level=1) -> 175.0
11      ),
12      baseline_dimensions_breakdown=Map(
13          Dimension(name=sm_code, value=AIR, group=delivery, level=1) ->
14              25.77777777777778,
15          Dimension(name=ca_state, value=MO, group=spatial, level=1) ->
16              79.66666666666667,
17          Dimension(name=ca_county, value=Kent County, group=spatial, level=2) ->
18              64.33333333333333,
19          Dimension(name=ca_state, value=WI, group=spatial, level=1) ->
20              78.88888888888889,
21          Dimension(name=ca_city, value=Spring Valley, group=spatial, level=3) ->
22              51.333333333333336,
23          Dimension(name=ca_state, value=MN, group=spatial, level=1) ->
24              104.22222222222223,
25          Dimension(name=ca_city, value=Oakland, group=spatial, level=3) ->
26              78.88888888888889,
27          Dimension(name=sm_code, value=SEA, group=delivery, level=1) ->
28              36.333333333333336,
29          Dimension(name=ca_county, value=Currituck County, group=spatial, level=2) ->
30              88.11111111111111,
31          Dimension(name=sm_code, value=BIKE, group=delivery, level=1) ->
32              5.111111111111111,
33          Dimension(name=sm_code, value=SURFACE, group=delivery, level=1) ->
34              15.555555555555555
35      ),
36      dimensions_hierarchy=Map(
37          Dimension(name=ca_county, value=Kent County, group=spatial, level=2) ->
38              Dimension(name=ca_state, value=MD, group=spatial, level=1),
39          Dimension(name=ca_city, value=Spring Valley, group=spatial, level=3) ->
40              Dimension(name=ca_county, value=Aleutians West Census Area, group=spatial, level=2),
41          Dimension(name=ca_city, value=Woodville, group=spatial, level=3) ->
42              Dimension(name=ca_county, value=Shannon County, group=spatial, level=2),
43          Dimension(name=ca_county, value=Sibley County, group=spatial, level=2) ->
44              Dimension(name=ca_state, value=MN, group=spatial, level=1),
45          Dimension(name=ca_city, value=Oakland, group=spatial, level=3) ->
46              Dimension(name=ca_county, value=Juneau County, group=spatial, level=2),
47          Dimension(name=ca_city, value=Stratford, group=spatial, level=3) ->
48              Dimension(name=ca_county, value=Currituck County, group=spatial, level=2),
49          Dimension(name=ca_county, value=Currituck County, group=spatial, level=2) ->
50              Dimension(name=ca_state, value=NC, group=spatial, level=1),
51          Dimension(name=ca_county, value=Shannon County, group=spatial, level=2) ->
52              Dimension(name=ca_state, value=MO, group=spatial, level=1),
53          Dimension(name=ca_county, value=Juneau County, group=spatial, level=2) ->
54              Dimension(name=ca_state, value=WI, group=spatial, level=1)
55      ),
56      records_in_baseline_offset=9
57  )
```

Fig. 7. AggregatedRecordsWBaseline example with truncated dimensions

5.3   Anomaly Detection Module

We describe this part very briefly, since the focus is on the RCA part and the material of ThirdEye's AD in Sect. 3 can be transferred in a straightforward manner or other state-of-the-art streaming point anomaly detection algorithms, e.g., from [34,48] as discussed in Sect. 2, can be integrated. In practice, each anomaly detector needs to implement the *AnomalyDetector* interface/trait to define (i) an *init* method that initializes the detector with a specification and (ii) a *runDetection* method, which receives a stream of events of type *DataStream[AggregatedRecordsWBaseline]* and outputs a stream of anomalies of type *DataStream[AnomalyEvent]*, where *AnomalyEvent* is the representation of an anomaly detected. The output stream of *AnomalyEvent* records has length equal to or less than the length of the input stream of *AggregatedRecordsWBaseline* records. Equal length means that every record in the input stream formed an anomaly. A *AggregatedRecordsWBaseline* record is included in every *AnomalyEvent* record as one of its attributes as it will be consumed by the RCA module in order to find the root cause of an anomaly.

5.4   Root Cause Analysis Module

To perform RCA based on any method, it is required to compare current value and dimension breakdown with a baseline. The baseline in our implementation is formed according to the *BaselineAggregator* defined previously. Each RCA method requires to receive as input a stream of type *DataStream[AnomalyEvent]*. The main goal of a RCA method is to calculate the *cost*, which represents the weight assigned to each dimension value in the dimension breakdown in order to quantify how important its change of value (compared to baseline) was and how much it affected the overall deviation of the metric that led to the anomaly detected. In our solution, we implement both the simpler flavor of contributors finder of StarTree ThirdEye community edition and the more advanced TC algorithm. The former is renamed to Hierarchy Contributors Finder. The *cost* in the case of Hierarchy Contributors Finder is the significance score as defined in the original ThirdEye implementation.

Change Measures and Result Representation. ThirdEye uses three different measures in order to quantify the change of a current dimension value compared to its baseline. More specifically, given the *current* and *baseline* values of a node along with the overall (root node) *currentTotal* and *baselineTotal* values, the three measures are calculated as follows:

$$valueChangePercentage = \frac{current - baseline}{baseline} \cdot 100 \tag{3}$$

$$contributionChangePercentage = \left(\frac{current}{currentTotal} - \frac{baseline}{baselineTotal} \right) \cdot 100 \tag{4}$$

$$contributionToOverallChangePercentage = \left(\frac{current - baseline}{|currentTotal - baselineTotal|} \right) \cdot 100 \quad (5)$$

To represent the results produced by a RCA method for a specific dimension in a summary, the *DimensionSummary* model is created, the schema of which is provided in Table 3.

Table 3. DimensionSummary schema

field	Scala type	description
Dimension	Dimension	The Dimension that summary concerns
currentValue	Double	Participation of Dimension in current value
baselineValue	Double	Participation of Dimension in baseline value
cost	Double	Significance score of this Dimension calculated by an RCA method
valueChangePercentage	Double	Measures the percentage change of current value, compared to baseline (Eq. (3))
contributionChangePercentage	Double	Measures the contribution percentage change compared to baseline (Eq. (4))
contributionToOverallChangePercentage	Double	Measures the overall percentage change of current compared to baseline (Eq. (5))

The analysis performed produces an instance of *RCAResult* model. The *RCAResult* model includes the id of the anomaly that was examined along with its *currentTotal* and *baselineTotal*. The actual explanation of the anomaly event is the list of *DimensionSummary* records that the *RCAResult* bears. This list is a distilled collection of the *Dimensions* that Contributors Finder methods identified as most important contributors to the anomaly detected, accompanied by change measures and ordered by *DimensionSummary.cost* in descending order. This model includes also the *dimensionGroup* property in order to declare which dimension groups this result concerns. In case a Contributors Finder algorithm produces a single result considering all different dimension groups, the value is set to "all", otherwise the name of the relevant dimension group is assigned.

Simple Contributors Finder. Simple Contributors Finder is the simplest RCA method that StarTree ThirdEye community edition offers [43] and it does not examine multiple dimensions jointly as DCA does through hierarchies. The implementation in the streaming setting, using Flink, is straightforward and fully corresponds to the implementation in the open source project [43] of StarTree ThirdEye community edition. It does not take into account the *Dimension* groups and the hierarchies formed as already mentioned.

More specifically, for each *Dimension* pair from the dimension breakdowns of both current and baseline found in *AnomalyEvent* records of the input *DataStream*, the change measures according to Eq. (4) and Eq. (5) are calculated provided that the overall change percentage contribution exceeds a threshold, e.g. 3%.

Finally, the method generates a *DimensionSummary* instance for each *Dimension* pair, to populate the dimension summaries list in the produced *RCAResult*. The list of *DimensionSummary* records is sorted by descending *cost*, having first discarded the records with *cost* equal to 0. The top k *DimensionSummary* records are kept where $k = summary\_size$, which is configured in the root cause analysis section of the application configuration.

Hierarchy Contributors Finder. Our implementation of the DCA algorithm, when examines a dimension value to compute its cost (i.e., the significance cost as defined in Sect. 3), it also considers its parent's values according to the group and the hierarchy it belongs to. To achieve this, there is an additional step added in this RCA process. When a stream of *AnomalyEvent* records is sent for analysis to the Hierarchical Contributors Finder, a flatMap method is applied in order to split a *AnomalyEvent* record into m new *AnomalyEvent* records, where m is the distinct number of *Dimension* groups observed in the set of *Dimension* records constructed by both current's and baseline's dimension breakdown.

This process is named *keyByDimensionGroup* and for each new *AnomalyRecord* of a group it generates, it populates the dimension breakdown with *Dimension* records only belonging to this specific group. Each new record has a format of (dimensionGroup, *AnomalyEvent*), where the dimension group is used as a key in order to perform the downstream RCA process per dimension group in parallel. The *AnomalyEvent* part of each new record is used to compute the cost according to the significance score, as described in Eq. (1). Additionally, the change measures from Eq. (3, 4, 5) are computed to populate the *DimensionSummary* explanatory model. Similarly to the *SimpleContibutorsCost* compute method, the cost is calculated only if the overall change percentage contribution threshold is reached. Additionally, in order to compute the parent-related quantities of the significance score formula, the dimensions hierarchy lookup map is used.

Table 4 provides an example case of a metric and some dimensions taken from the web sales cube as provided by the TPC-DS benchmark. The example has four dimensions that belong to two different groups, namely spatial and delivery. In Fig. 8, the same example is provided in a tree view. This is how a *AggregatedRecords* record would look like. In this example, Hierarchical Contributors Finder (i.e., our DCA implementation) would be executed separately for each group, meaning that the yellow tree forming the dimension breakdown of the spatial group would generate one *RCAResult* record and the purple tree would generate another one.

The *RCAResult* records produced by a Contributors Finder method are finally output back to a sink Kafka topic as shown in Fig. 4. This enables the proposed AD and RCA system to be integrated with other external systems, as the results of this process can be leveraged by downstream applications in an event driven fashion by employing Kafka consumers.

Table 4. Dimension breakdown example

Metric	Spatial			Delivery
ws_quantity	ca_city	ca_county	ca_state	sm_code
5	a1	b1	c1	BIKE
12	a2	b1	c1	AIR
8	a3	b2	c2	AIR
4	a3	b2	c2	BIKE
9	a4	b3	c1	AIR

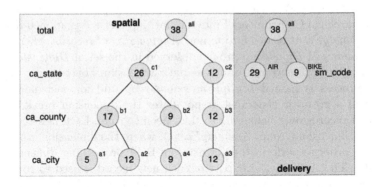

Fig. 8. Dimension breakdown example in tree view

6 Evaluation

In order to test the overall system, a time series dataset consisting of event records with at least one metric and some dimensions is required. The TPC-DS benchmark dataset [50] is used for this purpose, as it simulates a data warehouse of a business selling products across physical stores, website and catalogs. Sales data can be considered time series data as a completed purchase is a single event that takes place on a specific point in time. Also, sales data are accompanied with many metrics such as quantity sold, total order cost, shipping cost, vat and so on, and various dimensions such as payment method, product category, spatial information like city of delivery and more. The TPC-DS dataset can simulate a real world scenario in which a tool like the one proposed in this work would be helpful. Additionally, it is important to state that TPC-DS data generator provides a scaling factor in order to generate data of different scales, ranging from 1 GB to 100 TB. This provides the ability to stress test the scalability and performance of the solution. Finally, TPC-DS is a widely known and used publicly available benchmark dataset, so this work can be reproduced accurately and compared with other proposed solutions.

The dataset contains various fact tables for the different types of sales along with many dimension tables. In order to use TPC-DS, a Linux based Docker image was produced that is responsible for the generation of the dataset in the form of binary data files, one for each table of the data warehouse. As mentioned before, the data generation tool of TPC-DS provides a scale option in order to regulate the size of the generated dataset in GBs. We started with a dataset of 1GB, which results in a fact table of web sales with approximately 700 thousand records over a 5 years span and experimented up to scale 10 with approximately 7 million records over the same time period.

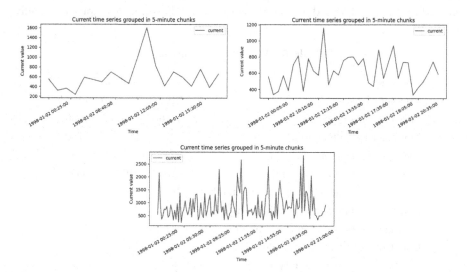

Fig. 9. Time series for sum of the ws_quantity metric for dataset with scale 1 (top-left), 2 (top-right) and 10 (bottom)

Regardless the scale of the TPD-DS dataset generated, the web sales cube consists of data with a time range [1998-01-02, 2003-01-02]. Datasets of higher scale are more dense, meaning that they have more data points in the same time range. Indicatively, in this section some analytics of the dataset in different scales is given in order to better understand the data used for the experiments. In Fig. 9, the time series of the sum of ws_quantity metric is provided for a dataset of scale 1,2 and 10. The sum aggregation is calculated over 5-min time chunks. The figure includes data only of the first day available in the dataset (1998-01-02) for reference, in order to showcase that a dataset of higher scale is more dense. Some additional measures to better understand the dataset these tests build upon are given in Table 5. Also, the cardinality of the dimensions we are interested to analyze in the RCA part is provided.

The machine used is equipped with a i7-8565U CPU, 16GB RAM and NVMe SSD. Regarding the setup for the test, 4 Flink Task Managers with 2 task slots each are used with default configurations as shown in the docker compose that is provided in the GitHub repository [18], while more details are in [19].

In Table 6 we present the total running times of our solution. Based on the results, it can be observed that the execution time grows linearly in the size of the input data. Note that the Metric Aggregation and Baseline Aggregation are calculated using a *windowAll* window method [7] that Flink provides, meaning that data are not partitioned and these two aggregation steps cannot be computed in parallel as chronological order of records needs to be maintained. Another notice is that despite the fact that the number of anomalies detected varies and, in the case of dataset of scale 10, there are about 120 times more anomalies detected compared to the scale 2 dataset (because the detection rule was tuned according to the scale 2 dataset), execution time does not grow proportionally,

Table 5. Web sales cube measures

Measure	scale 1	scale 2	scale 10
Number of records	719661	1438570	7195496
Max record count in a day	1166	2161	10308
Min record count in a day	72	205	1760
Average record count in a day	394.77	789.12	3947.06
Number of five-minute chunks	55037	101415	308997
Max metric value in a five-minute chunk	2957	3878	10614
Min metric value in a five-minute chunk	95	110	119
Average metric value in a five-minute chunk	661.18	716.4	1176.42
Count of distinct ca_state values	51	51	51
Count of distinct ca_county values	1846	1846	1846
Count of distinct ca_city values	631	723	965
Count of distinct sm_code values	4	4	4

Table 6. Execution time in seconds using different RCA methods and dataset scales

RCA method	scale 1	scale 2	scale 10
Number of anomalies detected	194	490	62015
Simple Contributors Finder	36 secs	61 secs	289 secs
Hierarchical Contributors Finder	34 secs	63 secs	305 secs

which indicates that the RCA process is not actually affected by the number of anomalies detected; to the contrary, it retains a constant processing rate. This is due to the fact that RCA is executed for all data, which may seem to be an unnecessary overprocessing but allows the provision of explanatory metadata to streaming statistics regardless of whether they constitute anomalies or not.

7   Conclusions

The main aim of this work is to present a fully operational research prototype that aims to transfer ThirdEye's RCA of detected anomalies in a streaming real-time setting. It is motivated by the fact that currently, there are no on-the-fly RCA solutions for AD. In this work, we describe all the pipeline steps in detail so that various current and window aggregates are computed in a principled manner to detect anomalies at any level of a data cube dimension hierarchy while providing RCA functionality. The whole system is implemented on top of Apache Kafka and Flink and can be easily extended by adding more AD and RCA algorithms, leveraging the existing models that describe the core entities of it. The system can further be extended by introducing different data stream sources and sinks other than Kafka and investigating additional forms of parallelism. The rich API of Flink provides many built-in connectors to achieve the former. The complete codebase is provided in [18].

References

1. Abuzaid, F., et al.: MacroBase. ACM Trans. Database Syst. **43**(4), 1–45 (2018). https://doi.org/10.1145/3276463
2. Angiulli, F., Fassetti, F.: Distance-based outlier queries in data streams: the novel task and algorithms. Data Min. Knowl. Disc. **20**(2), 290–324 (2010). https://doi.org/10.1007/s10618-009-0159-9
3. Apache: Apache flink: Stateful computations over data streams (2023). https://flink.apache.org/
4. Apache: Apache kafka (2023). https://kafka.apache.org/
5. Apache: Apache pinot: Realtime distributed olap datastore, designed to answer olap queries with low latency (2023). https://pinot.apache.org/
6. Apache: Apache zookeeper (2023). https://zookeeper.apache.org/
7. Apache: Flink, operators, windows (2023). https://nightlies.apache.org/flink/flink-docs-master/docs/dev/datastream/operators/windows/
8. Apache: Kafka connect overview (2023). https://kafka.apache.org/documentation/#connect
9. Basu, S., Meckesheimer, M.: Automatic outlier detection for time series: an application to sensor data. Knowl. Inf. Syst. **11**(2), 137–154 (2006). https://doi.org/10.1007/s10115-006-0026-6
10. Blázquez-García, A., Conde, A., Mori, U., Lozano, J.A.: A review on outlier/anomaly detection in time series data. ACM Comput. Surv. **54**(3), 1–33 (2021). https://doi.org/10.1145/3444690

11. Campos, G.O., et al.: On the evaluation of unsupervised outlier detection: measures, datasets, and an empirical study. Data Min. Knowl. Discov. **30**(4), 891–927 (2016)
12. Čampulová, M., Michálek, J., Mikuška, P., Bokal, D.: Nonparametric algorithm for identification of outliers in environmental data. J. Chemom. **32**(5), e2997 (2018). https://doi.org/10.1002/cem.2997
13. Carter, K.M., Streilein, W.W.: Probabilistic reasoning for streaming anomaly detection. In: 2012 IEEE Statistical Signal Processing Workshop (SSP). IEEE, August 2012. https://doi.org/10.1109/ssp.2012.6319708
14. Chang, Y.J.: Analyzing anomalies with thirdeye (2020). https://engineering.linkedin.com/blog/2020/analyzing-anomalies-with-thirdeye
15. Chen, J., Li, W., Lau, A., Cao, J., Wang, K.: Automated load curve data cleansing in power systems. IEEE Trans. Smart Grid **1**(2), 213–221 (2010). https://doi.org/10.1109/tsg.2010.2053052
16. Confluent: Kafka connect confluent documentation (2023). https://docs.confluent.io/platform/current/connect/index.html
17. Docker: Docker: Accelerated, containerized application development (2023). https://www.docker.com/
18. Flokas, Z.: Github repository: Zisisfl/online-anomaly-detection-root-cause-analysis (2023). https://github.com/ZisisFl/Online-Anomaly-Detection-Root-Cause-Analysis
19. Flokas, Z.: Online anomaly detection and root cause analysis (msc thesis) (2023). http://ikee.lib.auth.gr/record/347173/files/GRI-2023-38956.pdf
20. Fu, Y., Soman, C.: Real-time data infrastructure at uber. In: Proceedings of the 2021 International Conference on Management of Data. ACM, June 2021. https://doi.org/10.1145/3448016.3457552
21. Goldstein, M., Uchida, S.: A comparative evaluation of unsupervised anomaly detection algorithms for multivariate data. PLoS ONE **11**(4), e0152173 (2016)
22. Guidotti, R., Monreale, A., Ruggieri, S., Turini, F., Giannotti, F., Pedreschi, D.: A survey of methods for explaining black box models. ACM Comput. Surv. **51**(5), 1–42 (2018). https://doi.org/10.1145/3236009
23. Gupta, N., Eswaran, D., Shah, N., Akoglu, L., Faloutsos, C.: Beyond outlier detection: LookOut for pictorial explanation. In: Berlingerio, M., Bonchi, F., Gärtner, T., Hurley, N., Ifrim, G. (eds.) ECML PKDD 2018. LNCS (LNAI), vol. 11051, pp. 122–138. Springer, Cham (2019). https://doi.org/10.1007/978-3-030-10925-7_8
24. Han, J., Kamber, M., Pei, J.: Data Mining: Concepts and Techniques, 3rd edn. Morgan Kaufmann (2011). http://hanj.cs.illinois.edu/bk3/
25. Holešovský, J., Čampulová, M., Michálek, J.: Semiparametric outlier detection in nonstationary times series: case study for atmospheric pollution in Brno, Czech republic. Atmos. Pollut. Res. **9**(1), 27–36 (2018). https://doi.org/10.1016/j.apr.2017.06.005
26. Holt, C.C.: Forecasting seasonals and trends by exponentially weighted moving averages. Int. J. Forecast. **20**(1), 5–10 (2004). https://doi.org/10.1016/j.ijforecast.2003.09.015
27. Ishimtsev, V., Nazarov, I., Bernstein, A., Burnaev, E.: Conformal K-NN anomaly detector for univariate data streams (2017)
28. Jacob, V., Song, F., Stiegler, A., Rad, B., Diao, Y., Tatbul, N.: Exathlon: a benchmark for explainable anomaly detection over time series. Proc. VLDB Endow. **14**(11), 2613–2626 (2021)

29. Keller, F., Muller, E., Bohm, K.: HiCS: high contrast subspaces for density-based outlier ranking. In: 2012 IEEE 28th International Conference on Data Engineering. IEEE, April 2012. https://doi.org/10.1109/icde.2012.88
30. Keller, F., Müller, E., Wixler, A., Böhm, K.: Flexible and adaptive subspace search for outlier analysis. In: CIKM. ACM Press (2013). https://doi.org/10.1145/2505515.2505560
31. Ma, P., Ding, R., Han, S., Zhang, D.: Metainsight: automatic discovery of structured knowledge for exploratory data analysis. In: Li, G., Li, Z., Idreos, S., Srivastava, D. (eds.) SIGMOD '21: International Conference on Management of Data, Virtual Event, China, 20–25 June, 2021, pp. 1262–1274. ACM (2021)
32. Mehrang, S., Helander, E., Pavel, M., Chieh, A., Korhonen, I.: Outlier detection in weight time series of connected scales. In: 2015 IEEE International Conference on Bioinformatics and Biomedicine (BIBM). IEEE, November 2015. https://doi.org/10.1109/bibm.2015.7359896
33. Myrtakis, N., Christophides, V., Simon, E.: A comparative evaluation of anomaly explanation algorithms (2021). https://doi.org/10.5441/002/EDBT.2021.10
34. Ntroumpogiannis, A., Giannoulis, M., Myrtakis, N., Christophides, V., Simon, E., Tsamardinos, I.: A meta-level analysis of online anomaly detectors. VLDB J. **32**(4), 845–886 (2023)
35. Panjei, E., Gruenwald, L., Leal, E., Nguyen, C., Silvia, S.: A survey on outlier explanations. VLDB J. **31**(5), 977–1008 (2022)
36. Paparrizos, J., Kang, Y., Boniol, P., Tsay, R., Palpanas, T., Franklin, M.J.: TSB-UAD: an end-to-end benchmark suite for univariate time-series anomaly detection. Proc. VLDB Endow. **15**(8), 1697–1711 (2022). https://www.vldb.org/pvldb/vol15/p1697-paparrizos.pdf
37. Reddy, A., et al.: Using gaussian mixture models to detect outliers in seasonal univariate network traffic. In: 2017 IEEE Security and Privacy Workshops (SPW). IEEE, May 2017. https://doi.org/10.1109/spw.2017.9
38. Ribeiro, M.T., Singh, S., Guestrin, C.: "Why should i trust you?": explaining the predictions of any classifier (2016). https://doi.org/10.48550/ARXIV.1602.04938
39. Schmidl, S., Wenig, P., Papenbrock, T.: Anomaly detection in time series: A comprehensive evaluation 15(9), 1779–1797. https://doi.org/10.14778/3538598.3538602
40. Song, S., Zhang, A., Wang, J., Yu, P.S.: SCREEN. In: Proceedings of the 2015 ACM SIGMOD International Conference on Management of Data. ACM, May 2015. https://doi.org/10.1145/2723372.2723730
41. StarTree: Rca - top contributors (2023). https://dev.startree.ai/docs/startree-enterprise-edition/startree-thirdeye/concepts/rca-top-contributors
42. StarTree: Startree thirdeye (2023). https://dev.startree.ai/docs/startree-enterprise-edition/startree-thirdeye/
43. StarTree: Startree thirdeye community edition (2023). https://github.com/startreedata/thirdeye
44. StarTree: Startree thirdeye product features: Community vs enterprise edition (2023). https://dev.startree.ai/docs/startree-enterprise-edition/startree-thirdeye/ThirdEyeCommEdVsEntEdition
45. Taha, A., Hadi, A.S.: Anomaly detection methods for categorical data: a review. ACM Comput. Surv. **52**(2), 38:1–38:35 (2019)
46. ThirdEye: Thirdeye (2019). https://thirdeye.readthedocs.io/en/latest/
47. ThirdEye: Thirdeye archived github project (2022). https://github.com/project-thirdeye/thirdeye

48. Toliopoulos, T., Bellas, C., Gounaris, A., Papadopoulos, A.: PROUD: PaRallel OUtlier detection for streams. In: Proceedings of the 2020 ACM SIGMOD International Conference on Management of Data. ACM, May 2020. https://doi.org/10.1145/3318464.3384688

49. Toliopoulos, T., Gounaris, A.: Explainable distance-based outlier detection in data streams. IEEE Access **10**, 47921–47936 (2022). https://doi.org/10.1109/ACCESS.2022.3172345

50. TPC: Tpc-ds: Decision support benchmark (2023). https://www.tpc.org/tpcds/

51. Vinh, N.X., et al.: Discovering outlying aspects in large datasets. Data Min. Knowl. Disc. **30**(6), 1520–1555 (2016). https://doi.org/10.1007/s10618-016-0453-2

52. Winters, P.R.: Forecasting sales by exponentially weighted moving averages. Manage. Sci. **6**(3), 324–342 (1960). https://doi.org/10.1287/mnsc.6.3.324

53. Zhang, A., Song, S., Wang, J.: Sequential data cleaning. In: Proceedings of the 2016 International Conference on Management of Data. ACM, June 2016. https://doi.org/10.1145/2882903.2915233

54. Zhang, H., Diao, Y., Meliou, A.: Exstream: explaining anomalies in event stream monitoring. In: International Conference on Extending Database Technology (2017)

Characterization of the IPFS Public Network from DHT Requests

Bastien Confais[1]([⊠]), Benoît Parrein[2], Jérôme Lacan[3], and François Marques[1]

[1] Inatysco, 30 rue de l'Aiguillerie, 34000 Montpellier, France
{bastien.confais,francois.marques}@inatysco.fr
[2] Nantes Université, Polytech Nantes, rue Christian Pauc,
BP50609, 44306 Nantes, France
benoit.parrein@univ-nantes.fr
[3] ISAE Supaero, 10, Avenue Édouard-Belin, BP 54032, 31055 Toulouse, France
jerome.lacan@isae-supaero.fr

Abstract. Interplanetary File System (IPFS) is a file sharing network relying on a Distributed Hash Table (DHT) to locate data and a BitTorrent-like protocol to exchange blocks between the peers. However, in such a network, all nodes can access information about the files stored or accessed by others. In this article, we use these public pieces of information to try to characterize the IPFS network both in terms of nodes composing it and on the files stored in it. To that end, we set up an IPFS node connected to the public IPFS network and saved all the DHT requests forwarded through it. We show that nodes are mostly located in datacenters and not on the end-users' computers and therefore that files are often accessed through public gateways. We also show that most files are not replicated and are not accessed frequently (cold data) which can question us about the relevance of using IPFS in such use case scenarios.

Keywords: IPFS · Network monitoring · Kademlia DHT · Peer-to-Peer storage

1 Introduction

Distributed Storage solutions such as Google File System [10], often rely on a centralized metadata server and require to trust in storage nodes. Therefore, they can only be used within a single network provider or datacenter. To solve these major drawbacks, new alternatives such as Interplanetary File System (IPFS) [5], Sia [26] or Arweave [27] were developed in the last years. Some solutions are even built as an overlay of IPFS like Filecoin [6]. These solutions enable anyone, including non-trusted nodes to join the network. This is made possible thanks to replication of data and protocols checking the integrity of the returned data, preventing nodes from misbehaving. They also replace the centralized metadata server with a distributed structure such as a Distributed Hash Table (DHT) or

© The Author(s), under exclusive license to Springer-Verlag GmbH, DE, part of Springer Nature 2023
A. Hameurlain and A. M. Tjoa (Eds.): *Transactions on Large-Scale Data-and Knowledge-Centered Systems LV*, LNCS 14280, pp. 87–108, 2023.
https://doi.org/10.1007/978-3-662-68100-8_4

a blockchain. IPFS was developed in 2014 [5] and relies on a BitTorrent-like protocol [17] to exchange data between peers and on a Kademlia Distributed Hash Table [18] to locate the replicas.

In this paper, we propose a deep analysis of the public IPFS network, analyzing the nodes composing it and the pieces of data that are stored. We show that despite the need for the user to know the file identifier to retrieve the corresponding file, any node on the network can discover which files are stored and where, questioning the lack of privacy of using such networks. The remaining of the paper is organized as follows. Section 2, presents how IPFS works and lists some use cases of the use of IPFS that we can find in the literature. Section 3 presents the motivations of this work. Section 4 details how data was collected and Sect. 5 analyzes the results. Section 6 tries to take a step back and analyzes the consequences of the results before concluding in Sect. 7.

2   Interplanetary File System

IPFS [5] is a distributed storage solution relying on a BitTorrent-like protocol [17] to exchange data between the peers and on a Kademlia DHT [18] to locate the replicas. In the following, we explain how IPFS is working.

First, to join the network, a node connects to a bootstrap node in the DHT. The address of the bootstrap node is embedded in the source code of the software program as a default value for the node configuration. Once connected to the bootstrap node, the node sends DHT requests to build its routing table and determine which nodes it has to connect directly to. The neighboring nodes also send a replica of the DHT records to the new node, so it can be stored and served to other users. This is summarised in Fig. 1.

2.1   Writing Operation

When a user wants to write a new file, IPFS software program splits the file into different blocks of 256 KB that are stored on the local node. The block identifiers are the values of the hashes of the block contents. A Merkle tree is then built from these different blocks. The root hash becomes the file identifier (CID) and will be used to guarantee the integrity of the file (preventing the storage node to act maliciously) when retrieved. For each data block, the node creates a DHT record to indicate that it stores these blocks. This record is stored on the node in charge of managing the key of the block identifier. Therefore, the different records are spread within the IPFS network. Finally, the Merkle tree is also stored in the DHT. Figure 2 shows the full process of file creation.

2.2   Read Operation

To access a file, the user needs to know the value of the root hash of the Merkle tree corresponding to the file. A request is sent to the DHT to retrieve the Merkle tree. Then, a DHT request is sent for each block. This enables the node to get the

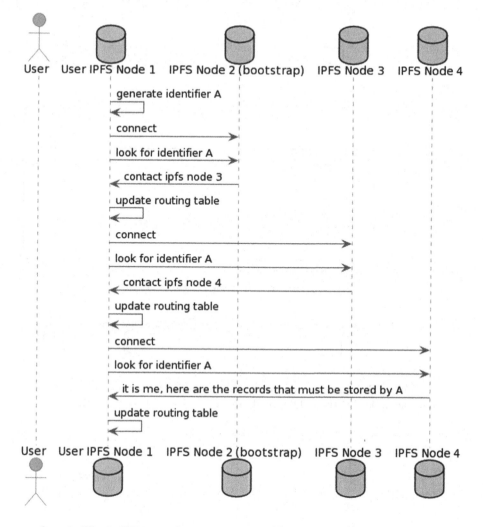

Fig. 1. Sequence diagram for the connection of a new node.

list of the available replicas for the block. Then, it can contact the nodes storing a replica to retrieve it. Finally, the copy of the retrieved blocks are stored on the local node, and the DHT is updated accordingly. This process is illustrated in Fig. 3.

Because of the DHT, many nodes are informed of the existence of the file, which is so not private. From a user's point of view, this can be counter-intuitive because the user can think that when the file identifier is not divulged, nobody except him can access the file.

2.3   Main Uses of IPFS

IPFS is currently used in different situations. A first use case is file sharing, that can benefit from the efficiency of the BitTorrent protocol and especially in video

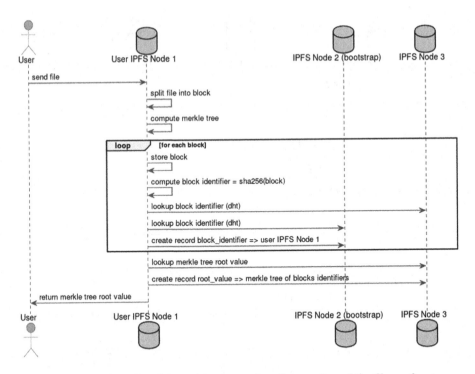

Fig. 2. Sequence diagram of the writing operation: the creation of the file on the storage node and the update of the DHT spreads among all nodes of the network

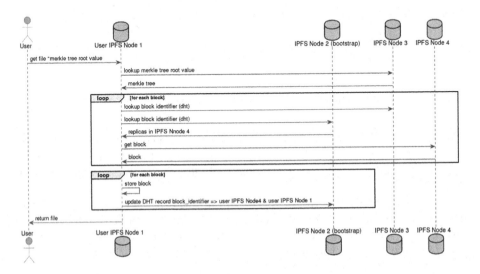

Fig. 3. Sequence diagram of the reading operation implying....

streaming application [9] or in the Internet of Things (IoT) to collect and store measurements [2,11]. Hosting a website [19], managing file revisions [14,20] or storing large container files are other use cases mentioned on the IPFS website[1]. However, coupled with a blockchain, IPFS can also be used for non-repudiation storage [24] providing a proof that a certain file exists and has been stored by a specific user. In practice, IPFS is used for traceability in supply chains [3] or for pieces of data produced by the devices of the IoT [16].

Monitoring IPFS software program has been done in many works. Shen *et al.* [22] or Abdullah *et al.* [1] evaluated the I/O performance of the program, but they deployed their own IPFS network and have the control over all nodes composing it. Public IPFS network has also been studied. For instance, it has been done in 2021 by Balduf *et al.* [4], Daniel *et al.* [8], Henningsen *et al.* [12] or even more recently by the organisation that develops the IPFS software program in Trautwein *et al.* [25]. We will present at the end of Sect. 4 the difference between their methodology and ours.

3   Motivations and Objectives

Before launching this study, we were interested in developing a software program that uses IPFS as a back-end for file exchanges between users. However, we quickly realised that the DHT could leak privacy information about the partners that are exchanging files but also that everybody could access the content of the file if it was not encrypted. The first goal of the paper is to demonstrate by experimentation that IPFS does not preserve the privacy of users. For that purpose, we would like to know if we can identify some peers that work together, that access the same files, indicating the movement of the user or two users that collaborate on a regular basis. The second goal of the paper is to gather as much information as possible about the nodes that are present in the network: what nodes are composing the network, when are they exchanging, but also on the files that are exchanged: their types, the number of replicas in order to characterize the network. The idea behind this is to try to determine the standard usage of IPFS, if people use it for its BitTorrent properties, or a storage like an alternative to Google Drive. We also want to make stable observations that are valid and reproducible. Therefore, we made the data collection from different nodes hosted on different providers.

4   Material and Methods

This study was made on the IPFS public networks that anybody can join and where all files are publicly available. We connected a IPFS node to the Internet for an entire year, from mid-December 2021 to January 2023 and saved the DHT requests that were forwarded by our node. The monitoring node was deployed on

---

[1] https://docs.ipfs.tech/concepts/usage-ideas-examples/.

the Google Kubernetes platform[2] [7] that could be reached using a public IPv4 address (Internet Protocol address) and by installing a full IPFS node (using the implementation developed in Golang[3], version 0.9.1). The node was deployed in region "europe-west3" located in Frankfurt, Germany. To validate the results and be assured that they are reproducible, we set up a second node on Amazon AWS. This node was located in Europe, Ireland.

We simply collected the DHT requests forwarded by our node by enabling the verbosity of the IPFS process.

Each saved record contains the date when the request was made, the file identifier (Content Identifier also known as "CID") and the identifier of the node which originally sent it ("PeerID"). In other words, a user using the node with the identifier "PeerID" is trying to locate the replicas of the file with the identifier "CID". We add that the DHT requests that were not resolved have been ignored. We used a Python script to extract more information from these records. For each line of log, we sent our own DHT request to the file identifier in order to retrieve all the peer identifiers that stores a replica. Then, we downloaded one of the replicas in order to identify the file type (MIME type). However, in order to preserve the privacy of users, we only stopped the download after the first few bytes (1000 bytes). For most of the file types can be determined with only the 24 first bytes. Nevertheless, some file types like Microsoft Office document require at least 512 bytes to be correctly identified[4]. Finally, for each peer identifier that stores a replica of the file, we resolved the identifier using the DHT in order to determine the corresponding IP address. With this piece of information, we could determine the network operator of the node and its geographical region thanks to the public databases published by the different Local Internet Registry (LIR) and aggregated in search trees by specific companies like MaxMind[5]. This whole process (capture of DHT requests and analysis of observed records) is summarised in Fig. 4. The source code of the developed scripts can be found at the following URL: https://github.com/Inatysco/IPFS-network-analysis. We also made our dataset available alongside with the source code but, we replaced all files identifiers, node identifiers and IP addresses with random strings.

4.1 Difference with Previous Studies

The global approach of exploiting the requests that we received from the DHT is not original and is similar to Balduf et al. [4], Henningsen et al. [12], Daniel et al. [8] Trautwein et al. [25]. For instance, Balduf et al. [4] showed that by intercepting the DHT requests, it is possible to determine the size of the network (the number of connected nodes), the activity levels, the structure and the content popularity distribution. Daniel et al. [8] made a similar work in 2022 by collecting passively the DHT requests to determine the network size and churn.

---

[2] https://cloud.google.com/.
[3] https://github.com/ipfs/kubo.
[4] https://www.garykessler.net/library/file_sigs.html.
[5] https://dev.maxmind.com/geoip/geolite2-free-geolocation-data?lang=en.

Henningsen *et al.* [12] also proposed to exploit the Kademlia DHT used by IPFS to crawl the network. The major difference with our work is that they use a software program called "Hydra" which is a DHT indexer that manages multiple "PeerID" in order to provide them a more global vision of what the DHT contains. Finally, Trautwein *et al.* [25] published an article in collaboration with the organisation that develops the IPFS software to analyze the distribution of requests made in the network. This last study has perhaps the approach which is the most similar to ours, by analysing nodes using their IP addresses and determining the number of requests received on a daily basis. However, all these studies focused on the nodes and none of them considered the files that were exchanged. The first originality of our approach is to send extra requests to complete the pieces of information that are collected directly from the DHT logs. For instance, we send the same request ourselves in order to receive the response and determine the place of all the replicas of the requested file. This led us to analyze the number of replicas or the position of the different replicas of a file in the network. Secondly, our study was made on a longer period of time (1 year) while Balduf *et al.* [4] was made on only 2 months and Daniel *et al.* [8] presents their results on only 10 days and Trautwein *et al.* [25] have a 7-months long analysis. This leads to detect much more nodes (three times more actually) in the public IPFS network than in these previous studies as we will see in the next section of results. Finally, our approach did not require using multiple "PeerID" and different positions in the DHT like in Henningsen *et al.* [12]. The long period of capture enabled us to see the environment changes and gave more opportunities for every node to contact ours. And secondly, the extra requests sent to complete the analysis were sent uniformly to the different peers, so that we could have contacts to the different zones of the DHT.

4.2   Ethical Question on Data Collection

Collecting such metrics is not uncommon for an analysis of a public network. It was already done to test peer-to-peer networks like BitTorrent [28] or Bitcoin [15].

Ethical concerns of such data collection from public networks has been studied by Small *et al.* [23] and by Partridge *et al.* [21]. Both studies show that the risk to identify the real person behind a node is small and that the possibility to exploit the collected pieces of information to cause harm to users is limited. They also distinguish active from passive data collection. When data is obtained passively and that by using the public network, users consent to see their pieces of information accessed by other nodes and they are free to leave the network. Generally speaking, the explicit consent of users of large-scale public networks as public IP network or P2P network is very difficult to obtain.

Our approach is semi-passive as we collect passively requests from the DHT but, we also send request to determine the type of stored files and the location of the replicas. However, these pieces of information are insufficient to identify the users (it would require us to know that a specific user stores a certain set of files).

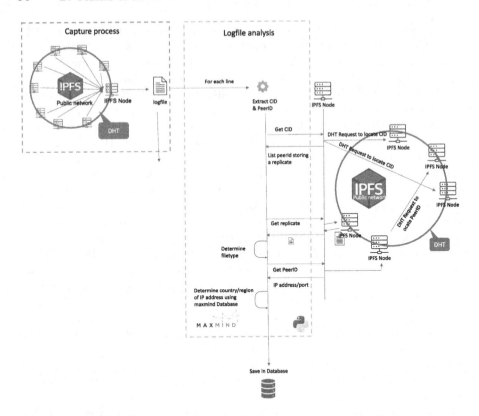

Fig. 4. Analysis process of the captured DHT requests.

If we compare our approach to previous papers, Badulf *et al.* [4] published a privacy policy on the website of their organisation to inform the users about the data collection. But there is no information about how IPFS users are concretely informed about their policy.

Trautwein *et al.* [25] added a paragraph in their paper to indicate that they have not tried to identify the real users behind the collected IP addresses. All the other works on IPFS mentioned in the previous section have not mentioned any information on ethics as recommended in [21].

5   Results

We collected 9 278 089 DHT requests that have been forwarded by our node (63 548 requests per day on average). In each request, we can see the identifier of the node sending it ("PeerID") and the "CID" (Content Identifier) of the file requested.

We also found that the node rebooted several times during the period, leading to observe few minutes without collecting any requests, but it also unexpectedly enabled it to change its position on the DHT because at each reboot, the node identifier of the node was modified.

The first general observation is that 13,5% of file exchanges are requested by only 10 nodes. These nodes probably correspond to public gateways, allowing any user to access files stored in IPFS, even if the user has not installed IPFS in his computer and does not actively participate in the network.

By observing the IP addresses associated to the node identifiers, we discovered in a second general observation that the network is composed of more than 147 000 nodes, that are mostly connected through providers operating in France, United States and Germany (detailed in Fig. 5). As a comparison, Balduf et al. [4] detect 48 000 different nodes in the period of their study. In the following, we will focus on the types of files that are stored in the network and their distribution among the nodes.

5.1   Type of the Requested Files

We collected approximately 230 000 file identifiers (CID) for which we could reach one of the replicas. After having downloaded the first 1000 bytes of each file, we could access to 7 074 files in plain text (TXT, JSON, ...), 5 199 PNG images and 3 953 PDF documents. All the types of file are summarised in Table 1.

Table 1. Types of files observed in the IPFS network (top 20)

MIME type	Number of occurrences	percentage
application/json	180 997	76%
application/octet-stream	35 012	14%
text/plain	7 074	2.9%
image/png	5 199	2.2%
application/pdf	3 953	1.7%
text/html	1 614	0.68%
image/jpeg	1 396	0.59%
application/gzip	812	0.34%
application/x-dosexec	366	0.15%
image/svg+xml	274	0.11%
image/gif	231	0.097%
video/mp4	227	0.095%
application/zip	162	0.06%
image/webp	116	0.049%
application/zlib	94	0.039%
text/x-java	78	0.033%
application/wasm	50	0.021%
audio/midi	39	0.016%
application/x-xz	39	0.016%
text/xml	39	0.016%

Three quarters of requests (76% exactly) are about JSON files. These mostly correspond to indexes: IPFS lets users send an entire hierarchy of files. All the identifiers of the files that are parts of the directory are added to a JSON structure and the JSON document is then stored in IPFS with its own file identifier. We have not specifically analysed the content of the directories (number of files, depth of the hierarchy, ...) but if a user accesses a file within a directory, a new DHT request is sent that could be captured and analyzed separately if our node receives the request for this resolution. In other words, when a user accesses a file like "QmXXXX/file.txt", a first request is sent to the DHT to resolve the CID of the directory: "QmXXXX". This returns a JSON object containing the filenames present in the directory, as well as the corresponding CID of the files. Then a second request is sent to the CID of the file to locate the replicas of it. In the worst case, if every file is accessed through a directory, we should see the directories representing 50% of all accesses. Because we see 76%, it means that either the position of our node led us to not capture the request for more than 2/3 of files access, or some users load the content of a directory without accessing any file. A third explanation and perhaps the most probable is that some files are under a deep hierarchy of folder (folderA/folderB/folderC/file.txt) leading to locate several JSON files before accessing the real file.

The 14% of total files that are binary files (application/octet-stream) correspond to files without recognisable header. It can be portions of a bigger file that have been split into several parts in order to enable users to perform random accesses, or it can be files that have been encrypted because the owner does not want anybody to read it. But most probably, these are tied to the fact that we only downloaded the first bytes of the files to preserve the privacy of users and we did not get sufficient information to determine the real type of the file. Therefore, it is difficult for us to understand what these files are about.

Although PDF or ZIP files can have an encrypted payload, most files with an identified type can be read by everyone. This illustrates and justifies the need of managing access permissions in such a network. Regarding the PDF files, we downloaded some of them and found that the immutability property of IPFS led some people to use it to store administrative documents such as proofs of delivery, invoices, etc. These private documents are made publicly available but probably without the consent of their owner. Medias like images (PNG, JPEG, WEBP) or videos (MP4) are stored on IPFS probably for optimising the distribution to several receivers[6]. In this case, the BitTorrent aspect of IPFS is probably what was looking for. We also note that many websites seem hosted using IPFS because we observe a lot of HTML files, but also Web assembly (WASM). These websites are usually reached through an IPFS gateway that can either be private or publicly accessible, like "ipfs.io".

The conclusion of this, is that most IPFS files are accessed through folders and that binary files from which we cannot determine the content, is the most common file type on the network. Then, we have a long tail with different kinds of files corresponding to different usages (PDF, images or websites).

---

[6] https://ipfs.video/.

5.2 Geolocalisation of the Nodes

We identified 147 003 node identifiers in the network but all the nodes have not the same importance. For instance, we identified 3 nodes announcing more than 2 000 IP addresses (mostly IPv6 addresses) that belong to a unique 64-bits long prefix network for each host. These nodes probably reconnect very often and announce a different address/port each time they reconnect.

In the same way, some IP addresses host different IPFS nodes, with different "peerID"s. This can correspond to nodes connected through NAT devices or IPFS nodes that reinstall their configuration on a regular basis. It can even be several IPFS daemons running on the same computer. A reason for this last case could be to improve the performance of IPFS as the daemon is only able to deal with one request at a time. Approximately 5 000 IP addresses host more than a single IPFS node.

From the node IP address, we determine a country to each node. As said in Sect. 4, countries are deduced from the Local Internet Registry (LIR) databases which can only reflect the administrative country of the network operator and not the real location of IPFS node. We luckily did not find any node with different IP addresses associated with different countries, therefore, if a node has several IP addresses, it is counted only once. However, if a node is restarted and has its identifier ("peerID") modified, it is counted as a new node.

Figure 5 shows the number of IPFS nodes that have been identified in each country. The figure shows that most IPFS nodes are located in the biggest economic zones of the world, which is not surprising. In Europe, most nodes are located in France (and Germany) and on the American continent, most nodes are located in the United States of America. The situation is different in Asia where nodes are more evenly distributed among the countries. But this can come from the fact that our node was located in Europe, prioritized connections to

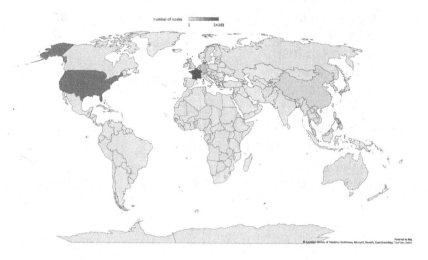

Fig. 5. Number of IPFS nodes per country.

nodes with low latency. Therefore, connection to nodes in Asia were made only when no other choice was possible leading to this map.

Table 2 and Table 3 show the distribution of nodes according to network operators. Like countries, the network operator of a node was determined from its IP address. Table 2 shows surprisingly that only 11% of the IPFS nodes are hosted on Google cloud, Microsoft Azure or Amazon AWS. However, Table 3 shows that the vast majority of the nodes are hosted in traditional datacenters and not in home networks. Providers like OVH, Packet, DigitalOcean hosts more than 45% of the IPFS instances. Only the last provider, T-Mobile is a provider for home networks. This questions the true decentralization of the IPFS network.

Table 2. Number of IPFS nodes hosted in famous cloud providers.

cloud provider	number of nodes identifiers	percentage
Amazon AWS	5 274	9.8%
Google	597	1.1%
Microsoft Azure	314	0.58%

In conclusion, the nodes are located in the most developed regions of the world which is not surprising. However, the more surprising result is that most of the node of the IPFS network are located in datacenters and not on the personal computers of the users.

Table 3. Number of IPFS nodes hosted in each Autonomous System (top 10).

network provider	number of nodes identifiers	percentage
OVH SAS (FR)	20 227	37%
PACKET (US)	5 737	10%
AMAZON-02 (US)	5 274	9.7%
AS-CHOOPA (US)	4 997	9.2%
DIGITALOCEAN-ASN (US)	3 218	5.9%
Contabo GmbH (DE)	2 995	5.5%
Verdina Ltd. (BLZ)	2 083	3.8%
Hetzner Online GmbH (DE)	1 950	3.6%
Hostkey B.v. (NLD)	1 895	3.5%
T-MOBILE-AS21928 (DE)	1 692	3.1%

5.3   Popularity of the Files

There are two ways to define the popularity of files:

– by looking at the replicas of each file - the file with the highest number of replicas is the most popular;
– by looking at the files that are requested the most in the captured DHT requests.

If we look at the number of replicas available for each file, we can build the Fig. 6. The figure shows a graph bar where each bar corresponds to a different file and the y-axis represents the number of replicas found for each file. We sorted the bars by the number of replicas.

It unsurprisingly shows a Zipf distribution with a very small number of popular files with a high number of replicas and a extra large number of files that are not popular and have fewer than 10 replicas. We can even see that 100 612 files (100612 files over 330638, approximately 30%) have less than 3 replicas.

Contrary to Balduf *et al.* [4] and as mentioned in Sect. 4, the requests that are not resolved are ignored. Therefore, we can observe a power-law distribution.

We observe that the file we found with the highest number of replicas has 11 515 replicas, and the second highest has 4 597 replicas. These files correspond to the "README" files of IPFS that are present by default on the nodes.

If we use the second criterion: by looking at the files requested in the intercepted requests, we can build Fig. 7. Like in the previous graph, each bar represents a file, and the height of the bar is the number of requests observed for the considered file. The bars are sorted according to their height.

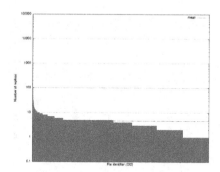
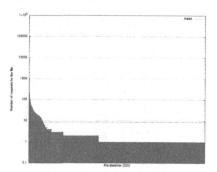

Fig. 6. Graph bar showing the number of replicas for each file (log scale in Y). Blue line represents the average number of replicas (average of 4.61). (Color figure online)

Fig. 7. Graph bar showing the number of requests for each file (log scale in Y). Blue line represents the average number of requests (average of 8.89). (Color figure online)

The first thing that we observe is that the average number of requests for each file is low (8.89 on average). If we look at the CID in the X-axis (not represented for readability), we observe that most queries are not about the files with most of the replicas (we do not observe any overlap in the 20 first CID of the two graphs in Fig. 6 and 7). In other words, the files with most of the replicas are not the most requested ones. This result seems strange because in IPFS each request on a new node leads to the creation of a new replica. This supposes that the popularity of files is changing over time, that many replicas are created when the file is popular, but then, the number of requests significantly reduces. This can explain why the files with the highest number of replicas are not the files with the most requests. We also note a bias in this figure. When a client requests a file to a node, the node looks for file replicas in the IPFS network by sending a DHT request. Then, a replica is downloaded, and a copy is created on the node that has received the request. Therefore, when a second client requests the same file to the node, no DHT request is sent in the network and can be captured. As a consequence, popular files that are requested by only a small set of IPFS nodes do not appear in the figure because DHT requests are not generated for each file access.

To conclude, the popularity of files shows a power-law distribution.

5.4   Activity in the Network

The fourth part of our analysis is about the use of IPFS network. We can show in Fig. 8 the number of requests intercepted for each day. We filtered the fact that nodes can send the same request several times within an interval of few

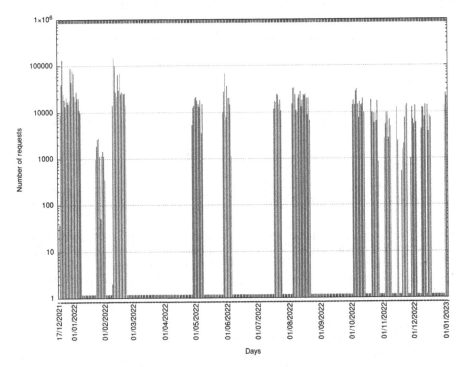

Fig. 8. Number of DHT requests received for each day.

seconds because it means that the node did not receive a response and retries to send the request.

Furthermore, we observe that the number of requests varies greatly with the days: 63 548 requests are sent on average each day, with a standard deviation almost equal to the average: 54 668 requests. However, we observe that, the number of requests at the end of January, was reduced significantly but increased again to reach the previous level on the 9th February.

5.5   Overhead of the DHT

A last question on the IPFS network is about its overhead in terms of network usage and energy consumption. In this part, we ignore the file transfer because if IPFS was replaced with a centralized solution, the same amount of data would have been exchanged. The only difference would have been that all flows originated from the same source rather than being distributed throughout the network. The real overhead of IPFS is the use of the DHT. DHT messages are composed of a message type on 4 bytes and the key that is looked for (on 38 bytes) Adding the IP and TCP headers, messages are 82 bytes long.

We received 9 278 089 requests. Because each request is 82 bytes long, we received $9278089 \times 82 = 760803298$ bytes in the period. From there, because the capture lasted 379 days 19 h 52 min, we can compute an average overhead throughput of 185 bps. However, the average throughput is not a relevant metric because, as it was shown in Fig. 8, requests are not received on a regular interval. If we look up at the peak of traffic on this figure, the maximum we observe is 37 687 requests on the 2nd February at 14:35. This corresponds to $37687 \times 82 = 3090334$ bytes for 60 s. Which is 412 044 bits per second (412 Kbps).

However, we saw in the previous section that more than 90% of nodes are hosted on a cloud computing platform and less than 10% are hosted on a computer connected to a consumer Internet Service Provider. Therefore, we can make the hypothesis that 90% of nodes are dedicated to IPFS and turned on only to run IPFS while 10% of nodes use the extra resources of computers used for other purposes. Nevertheless, all nodes are not necessarily physical ones, and the network is probably composed of virtual machines. The consequence is that running IPFS is not very different from running a cloud computing application in terms of energy consumption. The difference is mostly due to the fact that all nodes are not under the control of a single person with IPFS.

In conclusion, the overhead of the IPFS DHT in terms of network traffic is very low, but it is difficult to evaluate the number of dedicated physical nodes that have been added to the Internet to run the IPFS application.

5.6   Storage

In this last section, we study how the replicas are distributed among the IPFS nodes. We establish the list of peers, and for each file, we create a binary vector like (0 1 1) if we found that the file has a replica on the IPFS node 2 and IPFS node 3 but not on the IPFS node 1.

We therefore obtain a vector per file, and we can then compute correlations between files. A correlation of 1 indicates that the two files have their replicas on the same server, and a correlation of -1 indicates that the two files are stored on two complementary nodes.

Figure 9 shows a graph where each point is a file and the distance between two points is depending on the correlation value $distance = (1 - correlation)$. This distance means that two files that are stored on the same set of IPFS nodes are represented by two points at the same position (correlation $= 1$; distance $= 0$). On the contrary, two files that are stored on complementary nodes are represented far from each other (correlation $= -1$; distance $= 2$). We selected 200 files randomly.

It shows two clusters of files: one in the upper-right zone of the image and one in the bottom-left zone. The cluster of dots represented at the top of the figure corresponds to files that have only one replica on an IPFS server hosted on the network of DigitalOcean. The other cluster of dots, we can identify in the bottom of the figure, corresponds to the files that have a first replica on the node of the DigitalOcean network but also a second replica on a IPFS server hosted in the OVH network. All other files are files that have a unique placement of replicas: often a replica on a public gateway and another replica on the node of the user. These correspond to files that are not largely shared. This figure highlights different applications that use the IPFS network and have the tendency to place the file replicas on the same nodes.

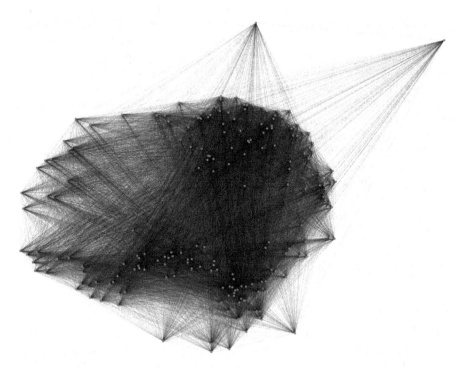

Fig. 9. Correlation between the storage of each file. Each dot represents a file and the edges between the nodes are depending on the servers the replicas are stored on.

5.7   Replication of Results on Amazon Web Services Platform (AWS)

In this last part, we try to determine if the observed results are reproducible. We run the same study by collecting DHT requests from a node hosted on Amazon Web Services[7]. This study was made over a shorter period of time: only 10 months from January to November 2022. Results show similar observations to those made using the node hosted on Google Cloud Platform in Table 1. We collected 593 451 DHT requests. We collected far fewer requests than on the Google node because of its latency and therefore probably a low score in the buckets of the Kademlia DHT.

File type distribution is similar to the distribution we observed on Google node. 75% of observed files are JSON objects and 18% are binary files as illustrated in Table 4.

Regarding the distribution of requests received, we observe the same Zipf distribution as we saw previously. Few files are requested many times and many others are requested only once. This is illustrated in Fig. 10. This short analysis with another node reinforces the observations we made over a longer period with the node hosted on Google Cloud Platform from Sect. 5.1 to Sect. 5.6.

6   Discussion

These results are surprising in the sense that IPFS is designed to share files efficiently with a large number of users. However, we observe that most of the files have few replicas and are therefore not shared with a large group of people. We can therefore wonder why these people use the IPFS solution that does not seem ideal because the BitTorrent-like protocol is not exploited to its full capabilities: we have not observed files with many replicas spread on a large set of nodes and we do not observe files becoming popular over time.

A possible reason is that IPFS is used for the use cases listed in Sect. 2.3 like non-repudiation storage [24] or traceability in supply chains [3]. IPFS seems to be more used for its immutability property that preserves files from modifications, allowing users to guarantee the integrity of the distributed files than for the file distribution around the globe. The distribution of accesses can also let us think that IPFS might be used by technical people who try it by creating a file, requesting it and leaving the network. In addition to this, we note that an IPFS gateway provides HTTP links to stored files, which can then be distributed easily among the users, making IPFS an alternative to services like drives (Dropbox, Onedrive, Google Drive) or File Transfer services (Wetransfer). Finally, as we observed many HTML files and images, we can tell that IPFS is used as a web hosting server. Websites are then accessed through IPFS gateways, achieving the scalability of web services with almost no extra cost (IPFS is acting like a Content Delivery Network).

---

[7] https://aws.amazon.com/.

Table 4. Types of files observed in the IPFS network from the AWS-hosted node (top 10)

MIME type	Number of occurrences	percentage
application/json	5 141	76%
application/octet-stream	1 222	18%
text/plain	168	2.5%
image/png	82	1.2%
image/jpeg	55	0.81%
text/html	24	0.35%
application/pdf	21	0.31%
application/gzip	16	0.24%
application/x-dosexec	12	0.18%
image/svg+xml	8	0.12%

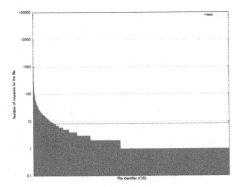

Fig. 10. Graph bar showing the number of requests for each file (log scale in Y). Blue line represents the average number of requests (8.22). (Color figure online)

A last remark is the concern about the need for security measures to preserve the users' privacy. In the current state of development, IPFS does not seem to give more guarantees to protect users privacy than the big cloud providers (GAFAM) do. It seems relatively easy for a malicious user to deploy many IPFS nodes in order to track most activities on the network. This tracking can be mitigated by spreading more replicas, even on nodes that do not access the piece of data. This prevents the DHT from being used when access to the data is made, because with IPFS, DHT requests are only sent when a local copy of the requested data is not found on the local node.

Also, a simple solution would be to encrypt both DHT records and files. The encryption key could be appended to the file identifier like "<CID>-<encryption key>". The user would still look for the CID in the DHT (the key would not be transmitted to the DHT) but the retrieved record is encrypted, and the key in the second part of the identifier must be known to decrypt it and to locate the replicates. Such a solution is not ideal because observing the requests in the DHT still leads to knowing the list of users that are accessing files, but the number of replicas, their location as well as the type of file could not be easily determined. Some articles [13,29] and the IPFS website list different projects that exploit content encryption[8] but most of them only encrypt data without considering the DHT.

Another solution would be to manage storage pools. The DHT would contain "pool1 managed on node1, node2 and node3". Then, the objects would be identified with the pool they belong to "<CID>-<pool>". In this way, the user locates the pool in the DHT, then directly contacts the nodes of the pool to get the replica. In this way, the DHT requests do not indicate the file the user is looking for, and observing the requests does not indicate how many files are stored in the network. The drawback of such a solution is that it makes the placement more difficult: many files must be stored together in order to avoid having a pool for each file, which recreates the problem we are trying to solve.

7   Conclusion

This study exploits the pieces of information publicly available on the IPFS network. These pieces of information led us to characterize the IPFS network by identifying the network operators through which, nodes are connected. It also enabled us to determine the popularity of files, the type of files exchanged, and the number of replicas of each file. We showed in Sect. 5.1 that in addition to the many PDF or image files stored, IPFS is also used as a hosting provider for websites. In Sect. 5.2, we showed that most nodes are from France and Germany in Europe and from the United States for the America continent. We then studied in Sect. 5.3 and Sect. 5.4 the frequencies of requests and their distribution along the days. We showed that a very small number of nodes send a lot of DHT requests and that most of the files are not popular, questioning us about the centralization of the network. In Sect. 5.5, we evaluated the overhead of the DHT of IPFS and concluded that it only consumes 185 bps which is very low even if peaks require more throughput. Before finishing, in Sect. 5.6, we identified in a subset of files two sets of files with their replicas placed on the same nodes. These files correspond to different applications relying on IPFS. These results have been obtained using an IPFS node hosted on Google Cloud Platform but were replicated on Amazon Web Services (AWS) in Sect. 5.7. In this situation, we can wonder what the interest is, in using IPFS for storing non-popular files because the main advantage of IPFS is the distribution of popular files. We considered several hypotheses, like non-repudiation storage or just because of the simplicity of sharing data compared to what cloud drives propose. We can also question the security of the IPFS protocol because, with few resources, it can be feasible for a malicious user to track many of the accesses performed.

---

[8] https://docs.ipfs.tech/concepts/privacy-and-encryption/#encryption.

We also note that this study has one major drawback: accessing files that have a replica stored on the IPFS node interrogated did not generate a DHT request and therefore could not be captured and analyzed. Evaluating the proportion of this kind of request is a bit hard to do as IPFS is a decentralized network.

Finally, this work can be continued by designing mechanisms to manage access permissions on files stored in IPFS or making improvements in the DHT in order to prevent the possibility of tracking the users' activity.

References

1. Abdullah Lajam, O., Ahmed Helmy, T.: Performance evaluation of IPFS in private networks. In: 2021 4th International Conference on Data Storage and Data Engineering. DSDE 2021, New York, NY, USA, pp. 77–84. Association for Computing Machinery (2021). https://doi.org/10.1145/3456146.3456159
2. Ali, M.S., Dolui, K., Antonelli, F.: IoT data privacy via blockchains and IPFS. In: Proceedings of the Seventh International Conference on the Internet of Things. IoT 2017, New York, NY, USA. Association for Computing Machinery (2017). https://doi.org/10.1145/3131542.3131563
3. Altmann, P., Abbasi, A.G., Schelén, O., Andersson, K., Alizadeh, M.: Creating a traceable product story in manufacturing supply chains using IPFS. In: 2020 IEEE 19th International Symposium on Network Computing and Applications (NCA), pp. 1–8 (2020)
4. Balduf, L., Henningsen, S., Florian, M., Rust, S., Scheuermann, B.: Monitoring data requests in decentralized data storage systems: a case study of IPFS (2021). https://arxiv.org/abs/2104.09202
5. Benet, J.: IPFS - Content Addressed, Versioned, P2P File System. Technical report, Protocol Labs, Inc. (2014). http://arxiv.org/abs/1407.3561
6. Benet, J., Greco, N.: Filecoin: A decentralized storage network. Technical report, Protocol Labs, Inc. (2018)
7. Bisong, E.: Containers and Google Kubernetes Engine. In: Building Machine Learning and Deep Learning Models on Google Cloud Platform, pp. 655–670. Apress, Berkeley, CA (2019). https://doi.org/10.1007/978-1-4842-4470-8_45
8. Daniel, E., Tschorsch, F.: Passively Measuring IPFS Churn and Network Size (2022). https://arxiv.org/abs/2205.14927
9. Doan, T.V., Pham, T.D., Oberprieler, M., Bajpai, V.: Measuring decentralized video streaming: a case study of DTube. In: 2020 IFIP Networking Conference (Networking), pp. 118–126 (2020)
10. Ghemawat, S., Gobioff, H., Leung, S.T.: The Google file system. SIGOPS Oper. Syst. Rev. **37**(5), 29–43 (2003). https://doi.org/10.1145/1165389.945450
11. Hasan, H.R., Salah, K., Yaqoob, I., Jayaraman, R., Pesic, S., Omar, M.: Trustworthy IoT data streaming using blockchain and IPFS. IEEE Access **10**, 17707–17721 (2022)
12. Henningsen, S., Rust, S., Florian, M., Scheuermann, B.: Crawling the IPFS network. In: 2020 IFIP Networking Conference (Networking), pp. 679–680 (2020)
13. Karapapas, C., Pittaras, I., Polyzos, G.C.: Fully decentralized trading games with evolvable characters using NFTs and IPFS. In: 2021 IFIP Networking Conference (IFIP Networking), pp. 1–2 (2021). https://doi.org/10.23919/IFIPNetworking52078.2021.9472196

14. Khatal, S., Rane, J., Patel, D., Patel, P., Busnel, Y.: FileShare: a blockchain and IPFS framework for secure file sharing and data provenance. In: Patnaik, S., Yang, X.-S., Sethi, I.K. (eds.) Advances in Machine Learning and Computational Intelligence. AIS, pp. 825–833. Springer, Singapore (2021). https://doi.org/10.1007/978-981-15-5243-4_79

15. Koshy, P., Koshy, D., McDaniel, P.: An analysis of anonymity in bitcoin using P2P network traffic. In: Christin, N., Safavi-Naini, R. (eds.) FC 2014. LNCS, vol. 8437, pp. 469–485. Springer, Heidelberg (2014). https://doi.org/10.1007/978-3-662-45472-5_30

16. Krejci, S., Sigwart, M., Schulte, S.: Blockchain- and IPFS-based data distribution for the internet of things. In: Brogi, A., Zimmermann, W., Kritikos, K. (eds.) ESOCC 2020. LNCS, vol. 12054, pp. 177–191. Springer, Cham (2020). https://doi.org/10.1007/978-3-030-44769-4_14

17. Legout, A., Urvoy-Keller, G., Michiardi, P.: Understanding BitTorrent: An Experimental Perspective. Technical report (2005). https://hal.inria.fr/inria-00000156

18. Maymounkov, P., Mazières, D.: Kademlia: a peer-to-peer information system based on the XOR metric. In: Druschel, P., Kaashoek, F., Rowstron, A. (eds.) IPTPS 2002. LNCS, vol. 2429, pp. 53–65. Springer, Heidelberg (2002). https://doi.org/10.1007/3-540-45748-8_5

19. Nguyen, T.-T., Do, B.-L.: A novel model using CDN, P2P, and IPFS for content delivery. In: Dang, T.K., Küng, J., Takizawa, M., Chung, T.M. (eds.) FDSE 2020. CCIS, vol. 1306, pp. 51–62. Springer, Singapore (2020). https://doi.org/10.1007/978-981-33-4370-2_4

20. Nizamuddin, N., Salah, K., Ajmal Azad, M., Arshad, J., Rehman, M.: Decentralized document version control using ethereum blockchain and IPFS. Comput. Electr. Eng. **76**, 183–197 (2019). https://www.sciencedirect.com/science/article/pii/S0045790618333093

21. Partridge, C., Allman, M.: Addressing ethical considerations in network measurement papers: abstract. In: Proceedings of the 2015 ACM SIGCOMM Workshop on Ethics in Networked Systems Research. NS Ethics 2015, New York, NY, USA, p. 33. Association for Computing Machinery (2015). https://doi.org/10.1145/2793013.2793014

22. Shen, J., Li, Y., Zhou, Y., Wang, X.: Understanding I/O performance of IPFS storage: a client's perspective. In: Proceedings of the International Symposium on Quality of Service. IWQoS 2019, New York, NY, USA. Association for Computing Machinery (2019). https://doi.org/10.1145/3326285.3329052

23. Small, N., Meneghello, J., Lee, K., Sabooniha, N., Schippers, R.: A discussion on the ethical issues in peer-to-peer network monitoring (2012)

24. Sun, J., Yao, X., Wang, S., Wu, Y.: Non-repudiation storage and access control scheme of insurance data based on blockchain in IPFS. IEEE Access **8**, 155145–155155 (2020)

25. Trautwein, D., et al.: Design and evaluation of IPFS: a storage layer for the decentralized web. In: Proceedings of the ACM SIGCOMM 2022 Conference. SIGCOMM 2022, New York, NY, USA, pp. 739–752. Association for Computing Machinery (2022). https://doi.org/10.1145/3544216.3544232

26. Vorick, D., Champine, L.: SIA: Simple Decentralized Storage. Technical report, NebulousLabs, Boston (2014)

27. Williams, S.A., Diordiiev, V., Berman, L.: Arweave: A Protocol for Economically Sustainable Information Permanence. Technical report, Minimum Spanning Technologies Limited (2019)

28. Wolchok, S., Halderman, J.A.: Crawling BitTorrent DHTs for fun and profit. In: Proceedings of the 4th USENIX Conference on Offensive Technologies. WOOT 2010, pp. 1–8. USENIX Association, USA (2010)
29. Zhou, C., Sun, G., You, X., Gu, Y.: A slice-based encryption scheme for IPFS. Int. J. Secure. Network. **18**(1), 42–51 (2023). https://doi.org/10.1504/IJSN.2023.129898. https://www.inderscienceonline.com/doi/abs/10.1504/IJSN.2023.129898

Customised Concept Weighting: A Neural Network Approach

Alaa Zreik[✉] and Zoubida Kedad[✉]

University of Versailles Saint-Quentin-En-Yvelines,
45 Avenue des Etats-Unis, Versailles 78000, France
alaa_zreik@hotmail.com, zoubida.kedad@uvsq.fr

Abstract. The aim of concept weighting in ontologies or in other data graphs is to characterise the importance of each concept in a specific domain, and to determine its selective power. This is particularly useful for data analysis tasks. Existing works on concept weighting mainly exploit either the graph's structure or the frequency of the concept in the data instances. These works provide concept weights independently form the considered analysis task. We argue that these weights should vary according to the targeted task, and we introduce a neural network based approach which computes concept weights using regression on a customised multi-layered structure. The loss function used in the approach is specified according to a given labelling of the elements in the considered dataset. In this paper, we present the principles of our weighting approach and we report on some experiments showing its effectiveness on real data extracted from the national library of France describing the documents' conservation histories.

Keywords: Concept Weighting · Neural Networks · Prediction

1 Introduction

Data analysis tasks aim at extracting meaningful knowledge from datasets containing elements described by a set of features. For example, in the healthcare field, an element could be a patient associated with its healthcare history composed of several features related to the prognosis, treatments, and hospitalisations. Recent works have focused on integrating domain specific semantic knowledge in the analysis process, with the goal of improving the results. The considered features may correspond to concepts which can be organised into a tree, highlighting the existing concept categories and the subsumption relations between them. Such tree represents some of the knowledge related to the considered domain.

Considering a dataset that contains elements represented by features, and an IS-A hierarchy representing the features and their relationships, such hierarchy can be used in different analysis stages. For example, while computing the

© The Author(s), under exclusive license to Springer-Verlag GmbH, DE, part of Springer Nature 2023
A. Hameurlain and A. M. Tjoa (Eds.): *Transactions on Large-Scale Data-and Knowledge-Centered Systems LV*, LNCS 14280, pp. 109–126, 2023.
https://doi.org/10.1007/978-3-662-68100-8_5

similarity between the elements, which is a core task in data analytics, the hierarchy could be used to identify matching features that correspond to concepts having common hypernym or equivalence relations. Furthermore, it is possible that some concepts are more important than others in a given context and that they should therefore play a significant role during similarity computation.

Among the proposed approaches for assigning a weight to the concepts in a hierarchy or a graph, some works have focused on the information content of the concepts. Sanchez [10] analyses the existing ontology-based information content computation methods, which are based either on the data and the occurrences of the concepts, or on the structure of the graph where many characteristics could be extracted such as the level of abstraction of the concept in the hierarchy, the number of hypernyms and hyponyms, or the number of leaf concepts. Resnik [9] proposed the first approach to use the information content in the computation of the similarity between two concepts C_i and C_j, and then it has been extended in [6,7] by using the result of the Resnik approach in addition to the information content of each concept separately. Graph weighting methods can be separated into extensional approaches and intensional ones [1]. The extensional methods rely on the data, while the intensional methods rely on the structure of the graph.

A significant limitation of the weighting methods is that they assign the same weights regardless of the analysis task, which is inappropriate for long-lived systems which might be used for different analysis tasks over time. Given the same data but for different analysis tasks and goals, the extensional methods will provide the same weights relying on the data. The same holds for the intensional methods, which rely on the tree structure: unless the tree structure changes, the resulting weights will not vary.

This paper introduces a novel concept weighting approach that takes into account a predefined labelling of the elements in the dataset, corresponding to a specific analysis requirement. The method first transforms the tree into a customised neural network where each node represents a concept. The edges between the nodes represent the relationships between the graph concepts. Based on regression, the neural network learns the edge weights that give the best separation of the categories of the elements. Finally, the weights of the neural network edges are used to extract the weights of the concepts. We have performed some experiments on a dataset describing ten thousand documents extracted from the database of the French National Library in order to show the efficiency of our approach.

This paper is organised as follows. In Sect. 2, we present a statement of our problem. Section 3 discusses the related works on concept weighting. Section 4 is devoted to our customised concept weighting approach. Section 5 presents our experiments, and finally, Sect. 6 provides a conclusion and some future works.

2   Problem Statement

Let us consider a dataset that contain elements $E = \{e_1, e_2, ...e_m\}$. Consider a set of features $F = \{f_1, f_2, ..., f_n\}$, where each element e_i is represented by a set f_{e_i} of k features describing this element, such that $f_{e_i} = \{f_{i_1}, f_{i_2}, ..., f_{i_k}\}$, with $f_{i_j} \in F$ for $1 \leq j \leq k$.

In addition, let us consider that the elements of E are grouped into a set of disjoints categories. Each element e_i is associated to a label l_{e_i} representing the category to which e_i belongs.

Considering the predefined categories, some features may have higher discriminative power than others. The discriminative power of a feature denotes its ability to distinguish between the categories. For example, a feature that appears in elements that belong to the same category has high discriminative power and should have a more significant impact on predicting the category than other features that frequently appear in elements belonging to different categories. In addition, the features in a given domain may share common characteristics; some features can be more generic than others and may correspond to the generalisation of a set of specific features in F.

In our work, we consider the taxonomy corresponding to the IS-A relationships between the concepts provided by a domain ontology. This taxonomy is a tree where the vertices represent the concepts, and where each edge represents an IS-A relationship between two concepts. Let $T.R$ be the set of edges in the tree, where each edge is of the form $<c_i, IS-A, c_j>$. c_i and c_j are two concepts in the tree, and the edge indicates that the concept c_j is a generalisation of c_i. The former is referred to as the child, while the latter is referred to as the parent. Each feature $f_i \in F$ is represented by a leaf concept c_i in T. We suppose that T contains several levels of abstraction, the most generic one being the root concept.

Except for one concept, which we will refer to as the $T.Root$ concept, we assume that all of the concepts in T have one parent. In addition, we define the leaf concepts $T.Leafs$, which contains all concepts that do not have a child concept. Figure 1 shows a tree representing a set of concepts, and to simplify the reference to the concepts, we represent each concept with a unique id. The tree contains five leaf concepts represented by $\{K, L, N, I, J\}$. For example, the concepts I and J are grouped into a generic concept G. K, L and M are grouped into H. The grouping continues to reach the root concept A.

Several research works have proposed approaches for weighting concepts in an ontology. These methods are either based on statistical computations or on the tree structure, which are fixed values that do not change if the labels of the elements are the same. Using these methods, the weights of the concepts would be the same regardless of the categories of the elements. The problem tackled in this paper can be stated as follows. Let us consider:

– (i) A dataset E composed of elements e_i, where each element e_i is described by a set of features in F.

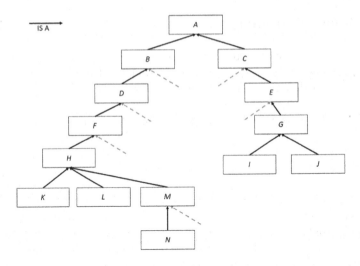

Fig. 1. Tree Representing a Concept Hierarchy

- (ii) A tree of concepts where each edge between two concepts represents an $IS - A$ relationship between them, and where the leaf concepts represent the features describing the elements in E.
- (iii) A partition of E where each subset correspond to a category of the elements in E such that each elements e_i in E is labelled by its category l_{e_i}.

Our problem is to determine the weight of each concept in the tree based on its discriminative power to distinguish between the different categories in E, taking into account the hierarchical links existing between concepts provided by the input tree.

This is different to the problem addressed by existing concept weighting approaches, in which weights are assigned independently from the specific task at hand. In other words, given two analysis tasks, such that the labels in the first task are l_1 and l_2, and the ones in the second task are l_3 and l_4, our weighting approach would provide two distinct sets of weights to the concepts, the first one corresponding to their ability to distinguish between l_1 and l_2, and the second one to their ability to distinguish between l_3 and l_4. To address this problem, we present a novel weighting approach that assigns weights to concepts organised as a tree based on their importance for a given partition of the considered dataset.

3 Related Works

Many works use the notion of information content (IC) to define the weights of the concepts corresponding to the nodes in a given graph [10]. Information content indicates the amount of information provided by a concept and its degree of generality. The computation of the information content of a concept is based on the inverse of its appearance probability in a given corpus [6,7,9], the number of hyponym concepts [11], or the depth of the concept [13] which is the distance to the root concept.

Some existing approaches give weights to the concepts independently of the notion of information content. Some are based on the data, and others are based only on the graph structure. The former are called extensional approaches, and the latter are called intensional approaches. This section presents the most frequently used graph weighting approaches.

Concept Frequency (CF) is an extensional approach based on the frequency of the concepts. Let us consider a set of elements, where each element has its annotation concepts, i.e. set of concepts describing it. Given a concept x, its frequency is the number of occurrences of x or one of its descendants in the elements' annotation concepts divided by the number of occurrences of all the concepts in the graph. The occurrences of a concept include all the occurrences of its hyponyms. Even if an abstract concept does not appear explicitly in an element's annotation concepts, the presence of one of its hyponyms indicates its existence. In formal terms, the CF weight of a concept x is defined as follows:

$$CF(x) = \frac{n(x^+)}{N}$$

where $n(x^+)$ is the number of occurrences of x or its hyponym concepts in the graph, and N is number of occurrences of all the concepts.

The Annotation Frequency approach (AF) is another extensional approach and was introduced in [3,4]. It is very similar to the CF approach in that both are based on concept occurrences. The difference between them is that the AF approach counts the number of element's annotation concepts that contain x or any hyponym while computing the weight of a concept x. If two hyponyms of x exist in an element's annotation concepts, they will be counted as one occurrence. On the other hand, the CF approach counts the number of occurrences regardless of whether they represent the same element. Finally, the CF approach divides by the number of occurrences of all the concepts, while the AF divides by the number of elements. It defines the weight of a concept x as follows:

$$AF(x) = \frac{|E_{x^+}|}{|E|}$$

where $|E_{x^+}|$ is the number of elements containing x or one of it hyponyms, and $|E|$ is the total number of elements.

The Top-Down Topology-based approach (TD) is a probabilistic and intensional approach based on the structure of the graph [1–3]. The approach starts by assigning a weight equal to one to the root concept and assumes a uniform probabilistic distribution along the $IS - A$ hierarchy; the weight of a concept x is its probability, which is computed as follows:

$$TD(x) = \frac{TD(parent(x))}{|children(parent(x))|}$$

where $|children(parent(x))|$ is the number of direct hyponyms of x's parent concept.

The semantic similarity presented in [1] is further extended into a parametric method for which one of the parameters is the approach used for computing the weights associated with the concepts of the ontology [8].

Another weighting approach is the Bayesian approach. It is intensional and based on the graph structure in addition to conditional probabilities. Starting from the specialisation relationship between a vertex x and its parent, the relationship indicates that the weight of x is influenced by the weight of its parent. The authors in [1] define the bayesian weight of a concept x by the probability that x is True (T):

$$w_b(x) = P(x = T)$$

This probability is related to the weight of the parent concept of x and the conditional probability $P(x = T|parent(x) = T)$. The weight $w_b(x)$ of a concept x is then computed based on the probability of its parent as follows:

$$w_b(x) = P(x = T|parent(x) = T)P(parent(x) = T) +$$
$$P(x = T|parent(x) = F)P(parent(x) = F)$$

where $P(x = T|parent(x) = F)$ is always equal to zero. Therefore, the simplified equation of $w_b(x)$ is the following:

$$w_b(x) = P(x = T|parent(x) = T)w_b(parent(x))$$

Both the extensional and intensional approaches were not designed to consider the analysis task at hand. By changing the task, they will provide the same concept weights because they only depend on the graph structure or on the occurrences of the concepts. For example, consider a task in the domain of document conservation–restoration, where a document can be in one of two following categories: $l_1 =$ "Out-of-order" and $l_2 =$ "Available" depending on its physical state. A document may have suffered multiple deteriorations, making it impossible for the readers to have access to it without inflicting more damages and accelerating the ageing process. The document in this case is said to be "Out-of-order". This could be due for example to the paper used to produce the document, which could be acidic paper. Assume that the considered analysis task is to determine the label of a given document. It is possible to find that the concept representing the observation of "acidic paper" for a document has a high discriminative power for this task. For another task with different categories, the "acidic paper" observation might not be as discriminative.

We provide in this paper a novel extensional weighting method based on neural networks that takes into account a given analysis task. The input of our approach is a set of elements represented by set of features, a concept tree representing the features and their semantic relationships, and a partition of the elements corresponding to different element labels. The method gives weights to the concepts representing their power to distinguish between the different categories. Therefore, by changing the labels of the elements, the proposed approach provides different weights.

4   Concept Weighting

Our approach introduces a novel concept weighting principle customised for a given analysis task considering some domain knowledge represented by an $IS - A$ hierarchy of concepts, and a set of labels defined for the elements of the considered dataset. Our aim is to find the best weights, i.e., the ones that give the highest accuracy for predicting the category of the elements in the dataset. The weight learning of the proposed approach is based on a customised neural network, and the importance of the concepts in the tree are learned using regression on a loss function defined according to the categories of the elements. Besides, our approach takes into account the relationships between the concepts during weight computation.

Each element e_i will be represented by a vector v_{e_i} to indicate which features exist in f_{e_i} to describe this element. In other words, a vector indicates which leaf concepts represent an element. The input of the neural network during the training phase is a set of vectors representing the elements as well as their labels.

The weighting is performed by transforming the tree into a customised neural network, and adapting its structure, i.e. the size of the layers, to vectorise the regression computation by adding empty nodes. Each node in the neural network represents a concept. Its position, i.e. layer, depends on the length of the path between the concept and the tree root $T.Root$. Regression is used to learn the weights in the neural network to minimise a predefined loss function. Finally, the learned weights in the neural network will be considered as the contribution of each concept in distinguishing between the input element categories and will be used to extract the concepts weights.

This section is organised as follows. In Sect. 4.1 we present the transformation of the tree to a neural network. Section 4.2 presents the normalisation of the neural network. Finally, Sect. 4.3 introduces the weights computation principle.

4.1   Neural Network Layers

The structure of the neural network (i.e. the number of layers, the size of each layer and the links between nodes) depends on the structure of the tree representing the concepts and their relationships. Therefore to build the customised neural network we should identify the following characteristics of the neural network.

Input Layer. The first layer in the neural network, i.e. the input layer, will represent the leaf concepts. Therefore, each element e_i will be represented by a vector v_{e_i} of size l, where l is the number of leaf concepts, i.e. size of $T.Leafs$, and each value in v_{e_i} is related to the number of occurrences of a specific concept in the feature representation of the element f_{e_i}.

Number of Layers. We define the tree depth $T.depth$ as the maximum path length between the root concept $T.Root$ and the leaf concepts $T.Leafs$, which will represent the number of layers in the neural network. Once the number of layers is calculated, the root concept is represented by a node in the output layer, and the distance between the root and every concept which is neither the root nor a leaf node is computed. Each concept c_i in the tree will be represented by a node denoted by n_{c_i} in the neural network. The layer L of a node n_{c_i} that represents the concept c_i is defined as $L = T.depth$-distance(root, c_i). For example, the longest path between the root and the leaves in Fig. 1 is the path between the concepts A and N, and it is equal to six.

Neural Network Links. Each parent concept is a generalisation of its children concepts. Therefore if a concept exists in the representation of an element, then implicitly, its parent also exist.

We therefore consider that the input of a node n_{c_i} is the output of all the nodes n_{c_j} such that c_i is a specialisation of c_j. In other words, the input of a node representing a concept c_i is the output of all the nodes representing the children of c_i. Using this principle, the core semantics of the specialisation relationship is preserved. The presence of a child concept in a set representing an element implies the presence of its parent. As a result, the value of a node representing a concept c_i is determined by the value of its children. Therefore the links between the nodes in the different layers depend on the relationships between the concepts in the tree. Figure 2 shows result of the transformation of the tree represented in Fig. 1 to a neural network where seven layers are created. The input layer L_0 contains five nodes, each of them representing a leaf concept. The first hidden layer in the neural network, L_1, contains one node representing the concept M, and the input of this node is the output of the node N in the input layer.

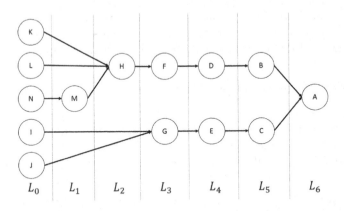

Fig. 2. Neural Network Layers

4.2   Normalising the Neural Network Structure

As mentioned before, each element in the training set will be represented by a vector of size l, which also corresponds to the number of nodes in the input layer.

Each value in a vector v_{e_i} representing an element e_i corresponds to a specific leaf concept in the tree. This value may represent the number of occurrences of the concept in e_i. The input of the neural network will be the set of vectors representing all the elements in the training set that will be used to learn the weights in the neural network. We recall in the sequel some of the well-known definitions related to neural networks such as input and weight vector [5]. We define an input vector as follows:

Definition 1. *Input Vector*
We define an input vector as a vector of l dimensions, where l is the number of nodes in the input layer, and we denote a single input vector as $a^{[0]}$.

For the example in Fig. 2, the input layer consists of five nodes; therefore, each element e_i will be represented by a five-dimensional vector. For instance, the vector representing the element e_i is $v_{e_i} = (1, 0, 0, 1, 1)$, which means that the concepts K, I and J exist in the element e_i and that these concepts occur once. The value of each node at the input layer will be equal to the corresponding value in the vector.

For each node in the neural network we define a weight vector as follows:

Definition 2. *Weight Vector*
A weight vector determines the input source of a node n either in a hidden layer or in the output layer. It will be used to calculate the node output. We denote a weight vector of a node n in layer L_i as w_n. The size of the weight vector will be equal to the number of nodes in layer L_{i-1}.

In our example, the node M in the first hidden layer should only take as input the third value from the previous layer because it is related only to the node representing the concept N. Therefore, the weight vector of the node M will be equal to $w_M = (0, 0, v_3, 0, 0)$ and the node output will be equal to $z_M = w_M.a^{[0]} + b_M$, where b_M is the bias term for M.

The node H at the second hidden layer L_2 takes as input values from different layers. H takes the output of the node M and of two nodes in the input layer, which are K and L. Therefore, it is difficult to define the size of the weight vector w_H. In addition, the nodes at layer L_3 have input values from different layers, and the length of their weight vectors will be different. Consequently, it is difficult to vectorise the calculations in each layer. As a solution, we propose to introduce empty nodes in the layers, defined as follows:

Definition 3. *Empty Node*
An empty node aims to fill a gap in the neural network where a node in a layer L_i should take an input from a node in a layer that precedes L_{i-1}. An empty node in a layer L_i takes its input from only one node in the layer L_{i-1} and will provide the same value as an input to a single node in the layer L_{i+1}.

Adding the empty nodes will ensure that each node at a layer L_i will take inputs only from nodes at layer L_{i-1} and the size of the weight vectors of all the nodes in a layer L_i is the same, and is equal to the number of nodes in the layer L_{i-1}.

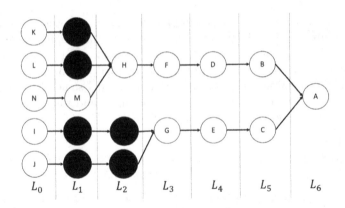

Fig. 3. Adding Empty Nodes to the Neural network Layers

Figure 3 shows the neural network after adding the empty nodes. At layer L_1, four empty nodes were added to provide the value of the nodes K, L, I and J from the input layer. The gap between nodes I and J with their parent was equal to two, i.e. the number of layers between L_0 and L_3. As a result, empty nodes for these two nodes were also added to the layer L_2.

Once the empty nodes are added, the computation in each layer can be vectorised. And a layer value vector that will be calculated is defined as follows:

Definition 4. *Layer's Value Vector*
The layer value vector depends on the node weight vectors, the input vectors and the node bias terms in the layer. We denote the values calculated in a layer L_i by $z^{[i]}$:

$$\underset{n \times 1}{z^{[i]}} = \underset{n \times m}{w^{[i]}} . \underset{m \times 1}{a^{[i-1]}} + \underset{n \times 1}{b^{[i]}} \tag{1}$$

$$\underset{n \times 1}{z^{[i]}} = \begin{pmatrix} w_1^{[i]T} \\ w_2^{[i]T} \\ \vdots \\ w_n^{[i]T} \end{pmatrix} \begin{pmatrix} a_1^{[i-1]} \\ a_2^{[i-1]} \\ \vdots \\ a_m^{[i-1]} \end{pmatrix} + \begin{pmatrix} b_1^{[i]} \\ b_2^{[i]} \\ \vdots \\ b_n^{[i]} \end{pmatrix} = \begin{pmatrix} z_1^{[i]} \\ z_2^{[i]} \\ \vdots \\ z_n^{[i]} \end{pmatrix}$$

where:

- $w^{[i]}$ *is a matrix of $n \times m$ dimensions containing the transpose of the weight vectors of the nodes in layer L_i, n is the number of nodes in layer L_i, and m is the number of nodes in layer L_{i-1}. Each row r in the matrix represents the weight vector of a node x in layer L_i.*
- $a^{[i-1]}$ *is a matrix of $m \times 1$ dimensions containing the output values of the layer L_{i-1} where $a^{[0]}$ is the input of the neural network.*
- $b^{[i]}$ *is a matrix of $n \times 1$ matrix containing the bias terms of the nodes in layer L_i.*

Finally, we define the output of a layer L_i as follows:

Definition 5. *Layer's Output*
The output of a layer depends on its values vector and the used activation function. We denote the output as act($z_{[i]}$), with act is the activation function used in the nodes of this layer.

In the example of Fig. 3, the layer L_2 contains three nodes, therefore $w^{[2]}$ is equal to:

$$
\begin{pmatrix}
w_{1,1}^{[2]} & w_{1,2}^{[2]} & w_{1,3}^{[2]} & 0 & 0 \\
0 & 0 & 0 & w_{2,4}^{[2]} & 0 \\
0 & 0 & 0 & 0 & w_{3,5}^{[2]}
\end{pmatrix}
$$

The values in $w^{[2]}$ indicate that the first node in layer L_2, which is H, takes input only from three nodes from L_1 with the weights $w_{1,1}^{[2]}$, $w_{1,2}^{[2]}$ and $w_{1,3}^{[2]}$ corresponding to the first, second and third nodes in L_1 respectively. Each empty node in the second layer takes its input from only one node from the previous layer.

4.3 Weight Learning

Weight learning starts by initialising the neural network parameters, which are the initial weight vectors of both the actual and the empty nodes, and the bias terms.In each iteration, forward and backward propagation [12] are performed.

Forward propagation is executed to calculate the prediction loss with the given parameters. Then backward propagation is performed to update the parameters so as to minimise the loss. Once the minimum loss is achieved, the final parameters will be used to compute the weights of the concepts in the tree.

Parameters Initialisation. Before weight computation, for each layer L_i, the bias terms $b^{[L_i]}$ and the weight vectors $w^{[L_i]}$ of the neural network nodes are initialised. $w^{[L_i]}$ is a n x m matrix where n is the number of nodes in layer L_i and m is the number of nodes in layer L_{i-1}. $w_s^{[L_i]}$ is the weight vector of the node at position s in layer L_i with $1 < s < n$, and $w_{s,p}^{[L_i]}$ represents the weight between this node and the node at position p in layer L_{i-1}. $b^{[L_i]}$ is a n x 1 matrix where each row contains the bias term of a node.

In the initialisation, we distinguish between the actual nodes and the empty nodes. For an actual node representing a concept c, the weights of the neural network links between this node and all the nodes representing the direct children of c in the previous layer will be initialised by assigning a random strictly positive value, and the other weights in the weight vector of this node will be equal to zero. The bias term will also be initialised to a random value.

Considering an empty node e_x at layer L_i having a predecessor node x from layer L_{i-1}, the weight between e_x and x will be initialised to 1 and the other weights in the weight vector of this node will be equal to zero. The bias term of this node will be equal to zero. For example, the weights $w^{[2]}$ of layer two in the neural network represented in Fig. 3 will be as follows:

$$\begin{pmatrix} r_{1,1} & r_{1,2} & r_{1,3} & 0 & 0 \\ 0 & 0 & 0 & 1 & 0 \\ 0 & 0 & 0 & 0 & 1 \end{pmatrix}$$

Forward Propagation. As mentioned in Sect. 2, our goal is to give weights to the concepts in the tree according to their importance in distinguishing between the categories. Each element e_i is represented by an l-dimensional vector v_{e_i}, where l is the number of nodes in the input layer. In addition, e_i has a label l_{e_i}. In a binary prediction, l_{e_i} takes one of two possible values. In our work we consider that $l_{e_i} \in \{0, 1\}$.

In forward propagation, the input of the neural network will be a set of vectors representing the training set denoted by $A^{[0]}$. The input is a n x m matrix, where n is the number of nodes and m is the number of elements in the training set.

The output of each layer L_i is $A^{[L_i]} = tanh(Z^{[L_i]})$ with $1 \le i \le TL - 1$ where TL is the total number of layers, and $Z^{[L_i]} = w^{[L_i]} A^{[L_{i-1}]} + b^{[L_i]}$.

For the output layer, we use the sigmoid activation function, $A^{[L_{TL}]} = \frac{1}{1+e^{Z^{[L_{TL}]}}}$, and the loss will be calculated using the logarithmic loss function:

$$\mathcal{L} = -\frac{1}{m} \sum_{x=1}^{m} (l_{e_x})(\log A_x^{[TL]}) + (1 - l_{e_x})(1 - \log A_x^{[TL]}) \tag{2}$$

Backward Propagation. Once the loss is calculated at the end of each forward propagation iteration, the weights and the bias terms for each layer are updated.

In the case of empty nodes, the weights should not be modified, and an empty node's output must always be equal to its input. Therefore, the weight vector of an empty node contains zeros except one value which is equal to one. In addition, the bias term of an empty node is equal to zero. Therefore, the weight vectors of empty nodes and their bias terms should not be modified.

The weight vector of each actual node is initialised depending on the relationship between the corresponding concepts in the tree, i.e., given two nodes n_s and n_p in layers L_i and L_{i-1} respectively, where n_s represents a concept c_s, and n_p represents a concept c_p, the weight between the two nodes should be always equal to zero if there is no $IS - A$ relationship between the two concepts in the tree.

Consequently, when updating the weight $w_{s,p}^{[L_i]}$ between nodes n_s and n_p representing the concepts c_u and c_v in layers L_i and L_{i-1} respectively, we distinguish between the following cases:

– Case 1: The nodes n_s and n_p are actual nodes and there is an $IS - A$ relationship between c_u and c_v. In this case, the weight $w_{s,p}^{[L_i]}$ will be updated depending on the learning rate α and the derivative of the loss with respect to $w_{s,p}^{[L_i]}$.

– Case 2: The node n_s is an actual node and n_p is an empty node that should pass the value to the node n_s. In the same way as for the first case, the weight $w_{s,p}^{[L_i]}$ will be updated depending on the learning rate α and the derivative of the loss with respect to $w_{s,p}^{[L_i]}$.

– Case 3: The nodes n_s and n_p are actual nodes and there is no a $IS - A$ relationship between c_u and c_v. In this case, the weight $w_{s,p}^{[L_i]}$ should be equal to zero.

– Case 4: The node n_s is an empty node. In this case the weight vector of the node n_p should not be modified.

In the first two cases, the equation to update the weight of the link between two nodes is the following:

$$w_{s,p}^{[L_i]} = w_{s,p}^{[L_i]} - \alpha \frac{\partial \mathcal{L}}{\partial w_{s,p}^{[L_i]}} \tag{3}$$

To generalise the weights update, we define for each layer L_i a n x m matrix called $Valid\_Input^{[L_i]}$, where n is the number of nodes in layer L_i and m is the number of nodes in layer L_{i-1}. This matrix is defined hereafter.

Definition 6. *Valid Input*
The valid input matrix indicates which values in the weight vectors could be changed according to the cases defined above.

The element $Valid\_Input_{s,p}^{[L_i]}$ is equal to one if the weight $w_{s,p}^{[L_i]}$ between the node at position s in layer L_i and the node at position p in layer L_{i-1} can be updated, and zero if it can not be updated.

The updated weight vector matrix $w^{[L_i]}$ of layer L_i is computed as follows:

$$w^{[L_i]} = w^{[L_i]} - \alpha \frac{\partial \mathcal{L}}{\partial w^{[L_i]}} \ Valid\_Input^{[L_i]} \tag{4}$$

The bias terms in the nodes are updatable for the actual nodes only. For this reason we define for each layer L_i a n x 1 matrix called $Nodes\_Type^{[L_i]}$ where n is the number of nodes in layer L_i, and where the element $Nodes\_Type_s^{[L_i]}$ is equal to one if the node at position s in the layer is an actual node and zero if it is an empty node. The equation to update the bias terms in layer L_i is the following:

$$b^{[L_i]} = b^{[L_i]} - \alpha \frac{\partial \mathcal{L}}{\partial b^{[L_i]}} \ Nodes\_Type^{[L_i]} \tag{5}$$

Concept Weights Calculation. Once the minimum loss is reached, we assume that the influence of the nodes on the output of the neural network represent their importance. In other words, the impact of a node representing a concept c_i on the prediction result, i.e. the neural network's final output, is the partial derivative of the final output with respect to this node's value. We use this principle to calculate the weights of the leaf concepts. The weights of the concepts at higher abstraction levels are calculated following a bottom-up approach depending on the children's weights. The impact of a node n_{c_i} representing a leaf concept c_i is defined as follows:

$$impact_{n_{c_i}} = \left| \frac{\partial A^{[TL]}}{\partial n_{c_i}} \right| \tag{6}$$

The weight of the concept c_i is the normalisation of the impact of the corresponding node:

$$weight_{c_i} = \frac{impact_{n_{c_i}} - \min_{c \in G.Leafs}(impact_{n_c})}{\max_{c \in G.Leafs}(impact_{n_c}) - \min_{c \in G.Leafs}(impact_{n_c})} \tag{7}$$

The weight of a non-leaf concept c_i is computed as the average of its children weights and defined as follows:

$$weight_{c_i} = AVG_{c \in hyponym(c_i)}(weight_c) \tag{8}$$

5   Experiments

The experiments presented in this section show that the proposed weighting method enables the accurate detection of the concepts that discriminate between the categories. The weights in the neural network are initialised randomly, and at each iteration we update the weights in the different layers according to a predefined learning rate, in order to minimise the loss results. The selection of the learning rate, the number of iterations and the type of the activation functions in the different layers has been done manually by testing different options and selecting the ones that give the best results, i.e. a minimum loss result. High weights are assigned to these concepts. Furthermore, we show that the analysis-based weighting method outperforms existing methods for detecting concepts with a high discriminative power by comparing the weighting results of our proposal with the ones of the state-of-the-art weighting methods.

The data used in the experiments is related to the Conservation-Restoration field and was extracted from the National Library of France (BnF) databases.

5.1   Data

In order to test our weighting approach, we have used an ontology in the domain of conservation-restoration called CRM_{BnF}, introduced in previous works [14,15]. This ontology contains concepts representing conservation–restoration events and their generalisation hierarchy.

We consider a set of documents $D = \{d_1, d_2, ..., d_n\}$, where each document d_i is represented by a set of concepts $Concepts_{d_i}$. Each concept is an event that has occurred in the conservation–restoration lifecycle of d_i. In addition, each document d_i is associated with a label l_{d_i} representing its physical state. We consider two categories of documents, "$Available$" and "$Out-of-order$". Our dataset is composed of 10000 documents. 70% of these documents belong to the "Available" category and 30% to the "Out-of-order" category. Each document is characterised by a set of features corresponding to the leaf nodes of the CRM_{BnF} ontology, and the total number of these leaf nodes (features) is 262.

5.2   Concept Weighting

The goal of this experiment is to assess the effectiveness of our weighting approach for the detection of the concepts with the highest discriminative power considering our two categories of documents. We first define the structure of the neural network considering the tree representing the concepts and their generalisation hierarchy, then this neural network is normalised, which leads to a neural network of 7 layers. The input layer contains 262 nodes representing the leaf concepts of CRM_{BnF}, and five hidden layers l_1, l_2, l_3, l_4 and l_5 containing 261, 249, 110, 26, 2 nodes respectively, and a output layer of one node representing the root concept of the ontology, i.e. $Event$. The weights are calculated and then normalised, ranging in the interval [0, 100], where 0 indicates that the concept has no discriminative power to distinguish between the two categories, and where 100 indicates that if the concept has the maximal discriminative power, and if this concept corresponds to the feature of a document, we can directly predict the document's label.

Table 1 shows ten concepts, their number of occurrences in the two document categories and their extracted weights. These concepts represent conservation-restoration events and have been specified by domain experts. The French names of these events are given in Table 1 but we will refer to them by their ID for the sake of clarity. Five of them, represented in green colour, have significantly high

Table 1. Analysis-Based Weighting Results

ID	Concepts	Occurrences in categories		Weight
		Available(/6962)	Out-of-order(/4641)	
3	Couture sur cahiers	687	28	77
72	Restauration reliure	59	0	100
4	Couture sur surjets	493	20	68
12	Couvrure Pleine toile	2425	123	37
165	COUV COINS Usure	15	127	37
152	MORSSUP Coupé	335	359	0.28
119	COUV PLATSUP Usure	503	577	0.12
144	COUV DOS Salissures	948	788	0.9
122	COUV Décollement	821	1227	0.65
147	COUV DOS Lacunes	800	1145	5.2

weights and the rest of the concepts, represented in red colour, have low weights. Concept 72 has the highest weight, with a value equals to 100. The analysis of its occurrences shows that this concept appears only in the "Available" category. In other words, if a document has this concept among its features, this indicates that the label of this document can be predicted to be "Available" with a 100% accuracy. The same holds for the other concepts with high weights; we can see that their distribution among the categories is not balanced, and they can be used to accurately predict the documents' labels. If we consider the concepts with low weights, we can see that their distribution among the different categories tends to be uniform, and therefore they have low discriminative power.

Table 2. Concepts Weights Using AF, CF, TD, Bayesian and Analysis-Based Methods

Concept's ID	Our Approach	AF	CF	TD	Bayesian
3	77	27	37	95	50
72	100	51	58	95	50
4	68	31	41	95	50
12	37	14	26	72	9
165	37	47	55	76	22
152	0.28	29	40	70	21
119	0.12	25	36	99	23
144	0.9	19	31	99	23
122	0.65	18	30	99	23
147	5.2	19	31	99	23

In Table 1, we have presented the occurrences of the concepts in the two categories in order to identify the relevant and the non-relevant concepts and we have shown that our approach assigns a high weight to the relevant concepts (in green colour) and a low weight to the non-relevant ones (in red colour). Table 2 shows the comparison of the weights provided by our approach with some of the methods presented in the related works (cf. Sect. 3). We can see that all the existing approaches do not succeed in identifying the relevant and non-relevant concepts in our dataset. For example, according to AF, concept 12 has a weight of 14 and concept 122 has a weight of 18. These concepts are given similar weights despite concept 12 being a relevant concept and concept 122 being a non-relevant one. We can also see that TD assigns high weights for all the concepts regardless of their relevance.

6   Conclusion

In this paper, we have proposed a novel concept weighting method based on a neural network. To this end, we have defined an approach for transforming a concept tree into a customised neural network. We have also proposed a learning process to derive the weights using regression on a predefined loss function that depends on the categories considered in the analysis task. The experiments have shown that our method outperforms the existing weighting methods to detect the discriminative power of concepts for distinguishing between the categories. The method gives high weights for the concepts that frequently appear only in one category and lower weights for those that frequently appear in more than one category.

In our future works, we will study the possibility of weights learning of concepts represented by a more complex graph, i.e. the relationships between the concepts are not only generalisations. In addition, we will study the case where the elements could be represented by concepts from different abstraction levels in the hierarchy.

We will also study in future work the impact of integrating these weights into a prediction model, where we will compare the results of the predictions, with and without the use of the calculated weights.

Another problem we will address in future research is the scalability of the proposed approach. In this paper, we conducted evaluations using a 7-layer tree, which determined the size of the neural network. However, when confronted with larger trees containing a high number of layers, the neural network's size would increase significantly, resulting in higher computation costs and complexity. Hence, as part of future research, we will explore the ways of scaling up our approach in order to process these larger trees more efficiently.

Acknowledgement. This work was supported by the Paris Seine Graduate School Humanities, Creation, Heritage, Investments for the future ANR-17-EURE-0021 - Foundation for Cultural Heritage Science.

References

1. De Nicola, A., Formica, A., Missikoff, M., Pourabbas, E., Taglino, F.: A comparative assessment of ontology weighting methods in semantic similarity search. In: ICAART (2), pp. 506–513 (2019)
2. Formica, A., Missikoff, M., Pourabbas, E., Taglino, F.: Weighted ontology for semantic search. In: Meersman, R., Tari, Z. (eds.) OTM 2008. LNCS, vol. 5332, pp. 1289–1303. Springer, Heidelberg (2008). https://doi.org/10.1007/978-3-540-88873-4_26
3. Formica, A., Missikoff, M., Pourabbas, E., Taglino, F.: Semantic search for enterprises competencies management. In: KEOD, pp. 183–192 (2010)
4. Formica, A., Missikoff, M., Pourabbas, E., Taglino, F.: Semantic search for matching user requests with profiled enterprises. Comput. Ind. **64**(3), 191–202 (2013)
5. Goodfellow, I., Bengio, Y., Courville, A.: Deep Learning. MIT Press, Cambridge (2016). http://www.deeplearningbook.org
6. Jiang, J.J., Conrath, D.W.: Semantic similarity based on corpus statistics and lexical taxonomy. arXiv preprint cmp-lg/9709008 (1997)
7. Lin, D., et al.: An information-theoretic definition of similarity. In: ICML, vol. 98, pp. 296–304 (1998)

8. Nicola, A.D., Formica, A., Missikoff, M., Pourabbas, E., Taglino, F.: A parametric similarity method: comparative experiments based on semantically annotated large datasets. J. Web Semant. **76**, 100773 (2023)
9. Resnik, P.: Using information content to evaluate semantic similarity in a taxonomy. arXiv preprint cmp-lg/9511007 (1995)
10. Sánchez, D., Batet, M., Isern, D.: Ontology-based information content computation. Knowl. Based Syst. **24**(2), 297–303 (2011)
11. Seco, N., Veale, T., Hayes, J.: An intrinsic information content metric for semantic similarity in wordnet. In: ECAI, vol. 16, p. 1089 (2004)
12. Svozil, D., Kvasnicka, V., Pospichal, J.: Introduction to multi-layer feed-forward neural networks. Chemom. Intell. Lab. Syst. **39**(1), 43–62 (1997)
13. Zhou, Z., Wang, Y., Gu, J.: A new model of information content for semantic similarity in wordnet. In: 2008 Second International Conference on Future Generation Communication and Networking Symposia, vol. 3, pp. 85–89. IEEE (2008)
14. Zreik, A., Kedad, Z.: Matching conservation-restoration trajectories: an ontology-based approach. In: RCIS, pp. 230–246 (2021)
15. Zreik, A., Kedad, Z.: Matching and analysing conservation-restoration trajectories. Data Knowl. Eng. **139**, 102015 (2022)

Author Index

A. Hameurlain and A. M. Tjoa (Eds.): *Transactions on Large-Scale Data-and Knowledge-Centered Systems LV*, LNCS 14280, p. 127, 2023.
https://doi.org/10.1007/978-3-662-68100-8